From Sovereignty to Guardianship

NEW TOPICS IN APPLIED PHILOSOPHY

Series editor: Kasper Lippert-Rasmussen

This series presents works of original research on practical issues that are not yet well covered by philosophy. The aim is not only to present work that meets high philosophical standards while being informed by a good understanding of relevant empirical matters, but also to open up new areas for philosophical exploration. The series will demonstrate the value and interest of practical issues for philosophy and vice versa.

ALSO PUBLISHED IN THE SERIES

Manipulation
Its Nature, Mechanisms, and Moral Status
Robert Noggle

Independence for Children
Matthew Clayton

What's Wrong with Lookism?
Personal Appearance, Discrimination, and Disadvantage
Andrew Mason

Disability Through the Lens of Justice
Jessica Begon

Exploitation as Domination
What Makes Capitalism Unjust
Nicholas Vrousalis

Sharing Territories
Overlapping Self-Determination and Resource Rights
Cara Nine

Spying Through a Glass Darkly
The Ethics of Espionage and Counter-Intelligence
Cécile Fabre

The Politics of Social Cohesion
Immigration, Community, and Justice
Nils Holtug

From Sovereignty to Guardianship

Governing Antarctica, Governing the World

ALEJANDRA MANCILLA

OXFORD
UNIVERSITY PRESS

OXFORD
UNIVERSITY PRESS

Great Clarendon Street, Oxford, OX2 6DP,
United Kingdom

Oxford University Press is a department of the University of Oxford.
It furthers the University's objective of excellence in research, scholarship,
and education by publishing worldwide. Oxford is a registered trade mark of
Oxford University Press in the UK and in certain other countries.

Published in the United States of America by Oxford University Press
546 Fifth Avenue, New York, NY 10036, United States of America

British Library Cataloguing in Publication Data
Data available

Library of Congress Control Number: 2026931810

ISBN 9780197902677

DOI: 10.1093/9780197902769.001.0001

Printed and bound by
CPI Group (UK) Ltd., Croydon, CR0 4YY

The manufacturer's authorized representative in the EU for product safety is
Oxford University Press España S.A. of Parque Empresarial San Fernando de Henares,
Avenida de Castilla, 2 – 28830 Madrid (www.oup.es/en or product.safety@oup.com).
OUP España S.A. also acts as importer into Spain of products made by the manufacturer.

MIX
Paper | Supporting
responsible forestry
FSC® C013604

For Alma, with love

Advanced Praise

'For more than 6 decades the Antarctic Treaty System has been a model of global governance, protecting the most pristine part of the Earth's surface from war and exploitation. These protections may not last long as the world increasingly eyes Antarctica's resources. In any case the existing legal framework cannot protect Antarctica from climate change, and its melting ice sheets threaten people around the world. Alejandra Mancilla's path-breaking, exciting and imaginative book begins the urgent task of reimagining how Antarctica should be governed in a way that protects the interests of all living things. While centered on political theory and philosophy, Mancilla's book is chockfull of history, stories, and anecdotes that will engage almost any reader.'

Dale Jamieson, Professor Emeritus of Environmental Studies and Philosophy at NYU, and author of *Reason in a Dark Time: Why the Struggle to Stop Climate Change Failed—and What It Means For Our Future.*

'With a clear-eyed assessment of the unsettling ecological and geopolitical contexts facing us, the philosopher Alejandra Mancilla finds reasons to be hopeful. By interrogating the governance regime for Antarctica, she detects ways of doing things that would recognise the needs of humans and non-humans in a potentially more respectful and just fashion. Brilliantly assembled, perfectly timed.'

Klaus Dodds, Hon Professor of Geopolitics Royal Holloway University of London and author of *The Antarctic: A Very Short Introduction* (OUP 2012)

'Engaging, imaginative, and forcefully argued. This book shows that anyone interested in the planet's future needs to pay serious attention to Antarctica.'

Chris Armstrong, Professor of Political Theory, University of Southampton and author of *A Blue New Deal: Why We Need A New Politics for the Ocean*

'This book is an important read for anyone interested in creating political institutions that can effectively safeguard our planet—and ourselves—now and into the future. By exploring the foundations, contradictions, and successes of Antarctic governance, Mancilla challenges us to throw out dated assumptions and imagine how to create political institutions that can preserve and protect life on earth.'

Claire Christian, Executive Director of the Secretariat of the Antarctic and Southern Ocean Coalition (ASOC)

Acknowledgements

This book has benefitted from questions and comments from audiences at the following venues: Tarragona Centre for Environmental Law Studies (CEDAT); King's College London; London School of Economics; University of Bristol; Center for Environment and Animal Protection (CEAP), New York University; University Center for Human Values, Princeton University; Arctic University of Norway; University of Cambridge; Friends of the Cam, Cambridge; Universidad Santiago de Compostela; Instituto de Ciencias del Patrimonio, Santiago de Compostela; KU Leuven; University of Oslo; Faculty of Law, Universidad de Valparaíso; Jornadas Antárticas, Universidad de Chile; Centro de Estudios Hemisféricos y Polares, Valparaíso; University of Bergen; University of Florida; Aarhus University; University of Potsdam; Queen's University Belfast; 'Climate and Territory' workshop, University of Waterloo; Nordic Network in Political Theory workshop, Oslo; Queen's University (Canada); 'Antarctica and Rights of Nature: Historical, Moral, and Legal Perspectives' workshop, Lysebu; University College Cork; University of Southampton; Universidad de Magallanes; NTNU; SCAR Conferences in Davos and Pucón; SC-HASS Conference, Lisbon; and Polar Law Symposium, Tromsø. A visiting fellowship at the Center for Environment and Animal Protection (CEAP), New York University, was very helpful to test the ideas in Chapters 5 and 6, and a research stay at the Scott Polar Research Institute, Cambridge University, helped me to find much of the material that appears in the book. It was a pleasure to get lost in that amazing library and to learn from other Antarctic enthusiasts during the daily morning tea—to which we were summoned by the ringing of the original bell of the *Terra Nova*, the ship in which Robert Falcon Scott sailed in his second expedition!

Although all the errors and stubbornness remain mine, I am indebted to many people for helping me to revise my ideas and for pushing me when more clarity and more argument were needed. For illuminating discussions along the years I would like to thank Anelio Aguayo, Kim Angell, Juan Carlos Aravena, Chris Armstrong, Elías Barticević, Kees Bastmeijer, Susana Borràs Pentinat, Duncan Bell, Chris Bertram, Megan Blomfield, Jean de Pomereu, Klaus Dodds, Luis Valentín Ferrada, Valentina Figueroa, Christel Fricke, Sarah Goff, Clare Heyward, Adrian Howkins, Kevin Hughes, Julia Jabour, Jennifer Jacquet, Dale

Jamieson, Carlos Joly, Avery Kolers, Elizabeth Leane, Consuelo León, Maria Marquès Banqué, Steve Marshall, Margaret Moore, Cara Nine, Paulina Ochoa-Espejo, Carmen Pavel, Tania Pivcević, Surabhi Ranganathan, Kerstin Reibold, Ricardo Roura, Pablo Ruiz, Juan Francisco Salazar, Christine Schwöbel-Patel, Ximena Senatore, Akiho Shibata, Peter Singer, Davor Vidas, and Leif Wenar. For written feedback, I am thankful to Patrik Baard, Alfonso Donoso, Andreas Føllesdal, Magnus Hoel, Jennifer Jacquet, Bjorn Sether Wastvedt, Yelena Yermakova, and four anonymous reviewers at Oxford University Press. I am also grateful to Hannah Winter, Oda Davanger, and Fabian Stenhaug for their excellent research assistance, and to Janet Martin-Nielsen for a thorough revision of the final manuscript. I owe a lot to some Antarcticans who encouraged me to pursue this line of research and generously shared their knowledge and answered many questions: Marcelo Leppe, Birgit Njåstad, José Retamales, and Peder Roberts have been unique sources of inspiration, each in their own way. The INACH community in Punta Arenas, where I visited in 2015 and 2016, provided the friendliness and warmth needed to dip into Antarctic research. And the Zoom meetings with the Antarctic Rights working group in the last 3 years have been a great incentive to move this project forward.

Special thanks go to the editors at OUP, Kasper Lippert-Rasmussen and April Peake, for making the process of publication smooth and for clarifying doubts along the way. I am also very grateful to Jean de Pomereu, who kindly let me use one of his evocative Antarctic artworks to illustrate the book cover.

Some of the chapters in the book draw from articles I have published elsewhere. Chapter 2 draws from 'The Moral Limits of Territorial Claims in Antarctica' (*Ethics and International Affairs*, 2018). Chapter 3 develops some of the ideas sketched in 'What Colonialism Tells Us About Antarctica, and What Antarctica Tells Us About Colonialism' (*Colonialism and Antarctica: Attitudes, Logics, Practices*, edited by Peder Roberts and Alejandra Mancilla, 2024) and in 'Decolonising Antarctica' (*Philosophies of Polar Law*, edited by Alan Hemmings and Dawid Bunikowski, 2021). Chapter 7 explores in further depth the ideas proposed in 'From Sovereignty to Guardianship in Ecoregions' (*Journal of Applied Philosophy*, 2023). The concept of the 'paradox of protection', which appears throughout the book, was initially proposed in 'The Paradox of Protecting Antarctica' (coauthored with Peder Roberts, *Aeon*, 2022). All these texts have been revised substantially.

Finally, this book could not have been written without the curiosity and awe for the world that my mother, Rita Drpić, instilled in me since I was little. I am indebted to her, and to my daughter Alma and socio Peder, for their unflinching love and support along the way.

Funding

The research for this book was supported by funding from the European Research Council (ERC) under the European Union's Horizon 2020 research and innovation programme (grant agreement no. 948964). Some of the ideas developed in the first part of the book can be traced back to the project 'Political Philosophy Looks to Antarctica' (funded by the Research Council of Norway) and, earlier still, to a workshop funded by the Norwegian Latin American Network (NorLARNet) at the Arctic University of Norway.

Contents

Illustrations

Map

Table

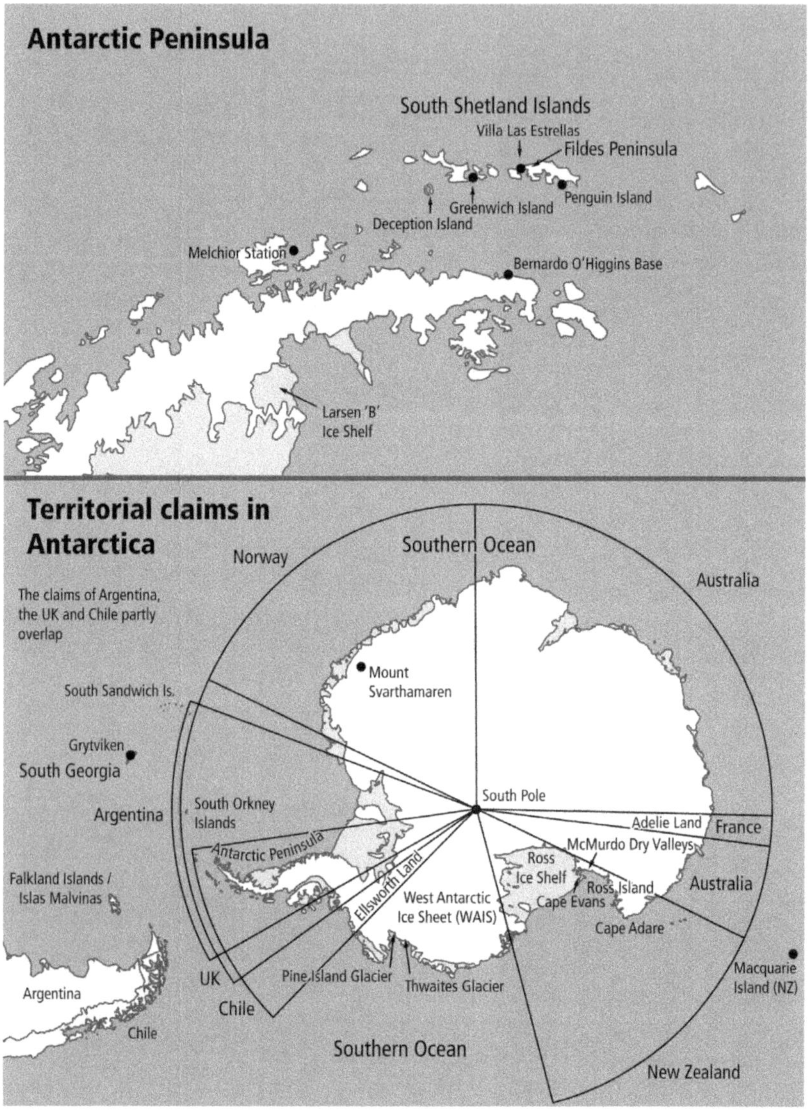

Top—Antarctic Peninsula and key locations mentioned in the book; Bottom—Territorial claims in Antarctica and key locations mentioned in the book. Maps by Hans van der Maarel, Red Geographics.

1
Introduction

The twin crises of climate change and biodiversity loss have shown humans our power to intervene and modify nature in ways that can end up threatening our own existence on earth. At the same time, and despite attempts to find technological fixes, they have shown us how limited our knowledge of and power to control the planet's processes and systems are. If there is any solution, or, rather, solutions to these challenges, they must be political. It is no accident, therefore, that scientists who have theorized the need to remain within our 'planetary boundaries' are advocating for global environmental governance to protect the 'planetary commons', and that researchers mapping the world's most valuable ecoregions are proposing the creation of a 'Global Safety Net' of protected areas going beyond national jurisdictions.[1]

While the need for developing environmentally sound principles of territorial governance to confront these challenges seems obvious, our political, legal, and institutional orders are steeped in two methodological biases that impede this development. On the one hand, methodological nationalism turns the stationary nation-state into the dominant (if not the only) territorial unit, effacing other territorialities. From this standpoint, the question is *how* to justify the rights of nation-states to, for example, control migration and dispose of the natural resources within their territories, rather than *whether* states should wield those rights in the first place. For everything that remains beyond national borders, meanwhile, the question has been *how* to justify a fair distribution of benefits among states, rather than *whether* there should be a distribution at all. On the other hand, methodological anthropocentrism makes humans into the dominant (and almost exclusive) units of political, legal, and

[1] See, respectively, Johan Rockström, Louis Kotzé, Svetlana Milutinović et al., 'The Planetary Commons: A New Paradigm for Safeguarding Earth-Regulating Systems in the Anthropocene', *Proceedings of the National Academy of Sciences* 121, no. 5 (2024): e2301531121, https://doi.org/10.1073/pnas.2301531121; E. Dinerstein, A. R. Joshi, C. Vynne et al., 'A "Global Safety Net" to Reverse Biodiversity Loss and Stabilize Earth's Climate', *Science Advances* 6, no. 36 (2020): eabb2824, https://doi.org/10.1126/sciadv.abb2824.

From Sovereignty to Guardianship. Alejandra Mancilla, Oxford University Press.
DOI: 10.1093/9780197902769.003.0001

institutional consideration. Accordingly, nature is seen as the background against which humans act and by and large as a resource at human disposal. The burden of proof is thus on those who wish to protect it rather than exploit it.[2]

And yet there is hope. There is one whole continent to which we can look for inspiration to create modes of territorial governance that go beyond these biases and respond more adequately to the environmental challenges that we face. Despite emerging from a colonial past, for over 60 years the Antarctic Treaty has ruled over Antarctica, preserving peace, science and—since the entry into force of the Protocol on Environmental Protection in 1998—the protection of the environment. Subject to neither the national regime of Permanent Sovereignty over Natural Resources nor the international regime of Common Heritage of Mankind, Antarctica has been jointly governed by a group of states with a focus on the protection rather than the exploitation of nature. The main argument of this book is that this regime of Joint Guardianship, where business has not been conducted as usual and where states have shown restraint, should serve as a blueprint for the governance of places beyond Antarctica—for example, ecoregions like the Amazon rainforest. At the same time, I argue that the Antarctic Treaty has not been wholly impervious to methodological nationalism and methodological anthropocentrism, even though it already contains some of the conceptual resources to move beyond them. If it were reinterpreted along these lines, this would result in a political system where not just humans, but also nature, are represented, and where the representatives are not just states, but also non-state actors. Furthermore, the book points to the 'paradox of protection' that affects Antarctica and other systemic commons and suggests a way to address it: to protect places like these, which are necessary for basic earth systems to function within certain limits, it is not enough to act only *in them*, but it is also necessary to act *beyond them*.[3] The take-home message is that governing Antarctica requires governing the world, and vice versa.

[2] Throughout the book I use the term 'nature' to designate the non-human natural world. At the same time, it should be borne in mind that a goal of the book is to remind humans that we are also part of nature and stand in continuity with it. Three sources for how to think differently about our relationship with nature and which have served as inspiration while writing this book are Jean Baptiste Morizot, *Manières d'être vivant* (Actes Sud 2020); Peter Godfrey-Smith, *Other Minds: The Octopus, the Sea, and the Deep Origins of Consciousness* (Farrar, Straus and Giroux, 2016); and Val Plumwood, *Environmental Culture: The Ecological Crisis of Reason* (Routledge, 2002).

[3] An initial development of the idea of the 'paradox of protection' can be found in Alejandra Mancilla and Peder Roberts, 'The Antarctic Paradox', Aeon, 27 May 2022, https://aeon.co/essays/how-do-we-solve-the-paradox-of-protection-in-antarctica.

1.1 Generating politics, illuminating politics

If there is one place on earth where human life can be as 'solitary, poor, nasty, brutish and short' as Thomas Hobbes painted it, that is Antarctica. Standard descriptions of the continent note that it is the coldest, windiest, driest, and highest on earth. The lowest temperature so far, recorded in August 2010 in East Antarctica, was –94.7 degrees Celsius, cold enough to send a person into a state of deep hypothermia in minutes.[4] Its katabatic winds, bringing the cold air from the heart of the continent to the coastal areas, beat any tropical storm, easily reaching more than 200 kilometres per hour. An ice desert, precipitations are on average 166 millimetres a year, very much like in Yemen, Oman, and Sudan.[5] Ninety percent of Earth's freshwater reserves are stored in Antarctica in the form of ice, making the continent's average height 2000 metres above sea level, reaching peaks of 4000 metres—peaks that only a few have dared to climb. To this it should be added that Antarctica is the most dynamically shaped continent and the most inaccessible: the sea ice makes it extend from 13 million square kilometres in summer to almost 20 in winter—as if roughly two Australias turned into three Australias every year, and back.[6] Surrounded by the Southern Ocean, its closest landmass is South America, the southernmost tip of which is nearly a thousand kilometres away.

It is no surprise, then, that Antarctica is the only continent without any Indigenous human population or permanent human settlements, and that it was the last to enter human history, after 30 million years of isolation. It was only two centuries ago that humans took the trouble to reach the Southern high latitudes, with a desire to explore as much as to exploit.[7] The near extinction of seals in the Arctic led seal hunters to move south, turning the

[4] Associated Press, 'Coldest Temperature Ever Recorded on Earth in Antarctica: –94.7°C (–135.8F)', *The Guardian*, 10 December 2013, https://www.theguardian.com/world/2013/dec/10/coldest-temperature-recorded-earth-antarctica-guinness-book.

[5] 'Annual Precipitation', Our World in Data, 2024, https://ourworldindata.org/grapher/average-precipitation-per-year.

[6] But this is rapidly changing; see Graham Readfearn, '"Everyone Should Be Concerned": Antarctic Sea Ice Reaches Lowest Levels Ever Recorded', *The Guardian*, 4 March 2023, https://www.theguardian.com/world/2023/mar/04/everyone-should-be-concerned-antarctic-sea-ice-reaches-lowest-levels-ever-recorded.

[7] This standard telling of Antarctica's origins may change in the future, if it is shown that it was Māori sailors who first reached the continent as early as the seventh century, as has recently been suggested. See Priscilla M. Wehi, Nigel J. Scott, Jacinta Beckwith et al., 'A Short Scan of Māori Journeys to Antarctica', *Journal of the Royal Society of New Zealand* 52, no. 5 (2022): 587–98, https://doi.org/10.1080/03036758.2021.1917633. For a critique, see Atholl Anderson, Tipene O'Regan, Puamiria Parata-Goodall, Michael Stevens, and Te Maire Tau, 'On the Improbability of Pre-European Polynesian Voyages to Antarctica: A Response to Priscilla Wehi and Colleagues', *Journal of the Royal Society of New Zealand* 52, no. 5 (2022): 599–605, https://doi.org/10.1080/03036758.2021.1973517.

sub-Antarctic islands and the coastal areas of the Antarctic Peninsula into a new 'resource frontier' where the animals were also driven close to extinction.[8] Then came the whale hunters, avid to collect blubber, an essential ingredient for the Industrial Revolution, needed to produce things as diverse as soap, candles, margarine, lubricants, and lamp fuel. Since the signature of the Antarctic Treaty in 1959, science became the predominant human activity on the continent. It is still only a few thousand who have encountered what Stephen Pyne called *The Ice*.[9]

Despite its geographic and climatic exceptionality, Antarctica plays a fundamental role in the earth's systems, influencing the patterns of atmospheric and oceanic currents, reflecting more sunlight than any other surface on the planet, and acting as a global thermostat. In times of climate change, Antarctica also becomes not only one of the most threatened places on earth, but also one of the most threatening: the ice discharge and sea-level rise resulting from the collapse of some glaciers in West Antarctica could have dramatic effects on coastal areas worldwide as soon as the coming decades.[10]

At the same time, despite its apparent aloofness from the general tides of human history, Antarctica is a territory 'created by politics, sustained by politics, and *generating* politics'.[11] It was to resolve the tension between seven territorial claimants (and two potential claimants) that the Antarctic Treaty was created and signed by twelve states on December 1st, 1959.[12] For over 60 years, this system of governance has ensured that a whole continent has been left non-militarized and non-nuclearized, to be used mainly for the furtherance of science. Moreover, it brought the continent under a unique territorial status, 'freezing' the sovereign claims so long as the Treaty is in force. Since the ratification of the Protocol on Environmental Protection (hereafter, Environmental Protocol), Antarctica also became—in a slogan—the most protected place on earth. Human activities, mainly science and tourism, were restricted to

[8] Adrian Howkins, *The Polar Regions: An Environmental History* (Polity, 2016), 47.

[9] Stephen J. Pyne, *The Ice: A Journey to Antarctica* (University of Iowa Press, 1986).

[10] Ted A. Scambos, R. E. Bell, R. B. Alley et al., 'How Much, How Fast?: A Science Review and Outlook for Research on the Instability of Antarctica's Thwaites Glacier in the 21st Century', *Global and Planetary Change* 153 (2017): 16–34; Ian Joughin, Benjamin E. Smith, and Brooke Medley, 'Marine Ice Sheet Collapse Potentially Under Way for the Thwaites Glacier Basin, West Antarctica', *Science* 344, no. 6185 (2014): 735–38, https://doi.org/10.1126/science.1249055.

[11] Alan D. Hemmings, Klaus Dodds, and Peder Roberts, 'Introduction: The Politics of Antarctica', in *Handbook on the Politics of Antarctica*, ed. Alan D. Hemmings, Klaus Dodds, and Peder Roberts (Edward Elgar Publishing, 2017), 1, their emphasis.

[12] The claimants were Argentina, Australia, Chile, France, New Zealand, Norway, and the United Kingdom. The two 'potential' claimants (who recognized no claims and left open the possibility of making future claims) were the United States and the Soviet Union. The other three signatories were Belgium, Japan, and South Africa.

minimize harm to its delicate environment, and a renewable ban of 50 years was set on mineral prospection and extraction (fishing, on the other hand, has always been allowed and is regulated by the Convention for the Conservation of Antarctic Marine Living Resources). Today, 58 states are members of the Antarctic Treaty, including all the major political and economic powers, representing more than two-thirds of humanity.[13]

This book starts from the claim that Antarctica is a territory that *generates* politics; a place where states, even if not for the most enlightened reasons, ended up committing themselves to a new form of territorial governance that should arguably serve as inspiration for the rest of the world. In Antarctica, as opposed to the rest of the world, the colonialist claims of seven countries over vast expanses of ice could never be legitimized, to the point where they had to be put on hold. This allowed for the creation of a unique territorial model, where nature is neither under the sovereignty of individual states nor under an international Common Heritage regime. Instead, member states exercise what I call Joint Guardianship through the Antarctic Treaty and its associated legal instruments. I argue that the Antarctic political imagination has responded creatively to new challenges for 60 years. In the 1980s, Antarctica was thought of as the last frontier for mineral prospection and extraction, and member states even drafted a Convention for the Regulation of Mineral Resource Activities in Antarctica (CRAMRA). However, growing global environmental awareness and criticism by non-member states gave rise instead to the Environmental Protocol. Signed in 1991 and ratified in 1998, the Protocol recognized the intrinsic value of Antarctica and turned it into a place where the protection of nature became the default.[14] The establishment of a Committee for Environmental Protection, moreover, gave a group of experts and scientists the task of advising the Treaty delegates on what should and should not be done on the continent. Despite geopolitical tensions, Antarctica has thus remained the only place left on earth where extraction and exploitation have not become the norm, with states showing an unusual dose of self-restraint.[15]

[13] Twenty-nine states are consultative parties, which meet annually to deliberate and create new regulations for the continent. Another twenty-nine states participate as non-consultative parties; that is, they can attend the meetings and voice their opinions but lack decision-making power. There are also observer and expert groups that can join the meetings, present their views, and advise the consultative parties—among them, the World Meteorological Organization, the Scientific Committee on Antarctic Research (SCAR), and the environmental NGO Antarctic and Southern Ocean Coalition (ASOC).

[14] The Environmental Protocol thus became one of the few international legal instruments to recognize the intrinsic value of a natural entity. See Antarctic Treaty System, 'Protocol on Environmental Protection to the Antarctic Treaty (1991)', Article 3, https://documents.ats.aq/recatt/Att006_e.pdf.

[15] Geopolitical tensions tend to make the news more often than political cooperation. See, for example, Elizabeth Buchanan, 'Great Power Competition Comes to Antarctica', *Foreign Affairs*, 18 March 2024, https://www.foreignaffairs.com/arctic-antarctic/great-power-competition-comes-antarct

But Antarctica is also a place that *illuminates* politics, uncovering blind spots and showing room for improvement. I suggest that the Antarctic Treaty—like other international law instruments—should overcome methodological anthropocentrism and methodological nationalism (more on these below) by including the representation of non-human Antarcticans and maybe even Antarctica itself, and by involving non-state actors in the decision-making process. Furthermore, because the protection of a place like Antarctica demands not just acting in it but also beyond it, those committed to its protection should not only act accordingly south of 60 degrees south, but also at home. Ultimately, Antarctic governance cannot be done by focusing on Antarctica only, and the converse is also true.

1.2 The Antarctic laboratory

In *The Polar Regions*, historian Adrian Howkins points to why doing an environmental history of the poles is a fruitful enterprise. First, he says, 'the strangeness of the Polar Regions offers new perspectives from which to view familiar subjects. Second, the role of the environment . . . cannot be ignored'.[16] Echoing Howkins, I suggest two reasons why looking at Antarctica may give us hints about how to conceive of territorial governance elsewhere, and especially in those places that, like Antarctica, serve a global function in maintaining the relative stability of our earth systems.[17] Natural scientists deem Antarctica to

ica and Anne-Marie Brady, *China as a Great Polar Power* (Cambridge University Press, 2017). That said, it is true that the Antarctic Treaty System has been moving very slowly in the last years and, as Alan Hemmings puts it, it has become 'increasingly dependent upon only hortatory Resolutions and assorted declarations reaffirming past legal commitments' (Alan D. Hemmings, 'Realising Rights of Nature Across and Beneath the Southern Ocean', *The Polar Journal* 14, no. 2 (2024): 409–26. https://doi.org/10.1080/2154896X.2024.2415826). Moreover, growing activities like tourism pose a challenge to the system. As Kees Bastmeijer et al. note, 'Under the legal design of the Antarctic Treaty and Protocol, Antarctica is open to peaceful use by all states and their nationals, except for activities that are explicitly prohibited or that are contrary to the principles or purposes of the Treaty or contrary to the Protocol' (Kees Bastmeijer, Akiho Shibata, Imme Steinhage, Luis Valentín Ferrada, and Evan T. Bloom, 'Regulating Antarctic Tourism: The Challenge of Consensus-Based Decision Making', *American Journal of International Law* 117, no. 4 (2023): 651–76, on 668, https://doi.org/10.1017/ajil.2023.34). In practice, this means that many activities that are potentially damaging for the environment can be carried out so long as there is no consensus among the consultative parties to explicitly prohibit them.

[16] Howkins, *Polar Regions*, 13.

[17] One may ask here why Antarctica only, rather than the polar regions in general (following Howkins), should be used as case study and testing grounds in this book. After all, they are both extreme environments that are key for the regulation of earth's basic systems; they have a similar history of exploitation of natural resources; and they are similarly affected by climate change. However, the differences between them are more important than their similarities. While the Antarctic is a continent representing 10 per cent of the earth's landmass, the Arctic is an ocean partially covered by ice. While Antarctica has never been inhabited by a permanent human population, the Arctic regions have been

be the greatest open-air laboratory on earth, providing privileged conditions for the study of geology, glaciology, and marine biology, among other disciplines. Similarly, I suggest that Antarctica should also be used as a laboratory to rethink territorial governance and its associated concepts.

First, in terms of offering a new perspective from which to examine familiar subjects (paraphrasing Howkins), Antarctica provides a unique entry point to think about topics like colonialism, political representation, and environmental protection. Colonialism, generally condemned as the unjust subjugation of one group of people by another and including features like domination, cultural imposition, displacement, and exploitation, seems like an odd category to be applied in a place with no native human population.[18] Yet, studying how colonialism manifested itself in Antarctica helps to isolate certain features that go unnoticed in lower latitudes. As I argue in Chapter 3, the unilateral hoarding of land by the seven claimants shows a morally problematic aspect of colonialism, even though it did not directly affect any humans. Regarding political representation, Antarctica again works like a laboratory, raising the following question, discussed in Chapters 4–6: How should the political community and its representatives be constituted in a place with no humans, but which has been brought under human governance? Finally, when it comes to environmental protection, Antarctica helps us to see the impossibility of trying to protect a place only by acting *in situ*. Like few places on earth, Antarctica exemplifies 'the paradox of protection'; namely, that to protect a place it is not enough to take action *within it*, but it is also necessary to take action *beyond it*. This points to the need for cooperation between actors for effective and adequate environmental governance.

Second, looking at territory through an Antarctic lens makes it clear that 'the environment' (or 'nature', as I call it) cannot be ignored. Even more strongly, it must be considered from the outset. So far, humans, their norms, and institutions have been the beginning and end of territorial governance,

inhabited by humans for millennia. Politically, moreover, Arctic lands and a big part of Arctic waters are partitioned between states, while Antarctica's territorial status remains unique and creates a system of coordination between states unseen anywhere else. In terms of protection, finally, there is nothing equivalent to the Environmental Protocol in the Arctic. These differences, however, should not detract us from thinking that some Antarctic values and principles could be fruitfully applied to the Arctic.

[18] For normative assessments of colonialism for which the subjugation of humans is an essential part, see Lea Ypi, 'What's Wrong with Colonialism', *Philosophy & Public Affairs* 41, no. 2 (2013): 158–91, https://doi.org/10.1111/papa.12014; Laura Valentini, 'On the Distinctive Procedural Wrong of Colonialism', *Philosophy & Public Affairs* 43, no. 4 (2015): 312–31, https://doi.org/10.1111/papa.12057; Margaret Moore, 'Justice and Colonialism', *Philosophy Compass* 11, no. 8 (2016): 447–61, https://doi.org/10.1111/phc3.12337; and Vittorio Bufacchi, 'Colonialism, Injustice, and Arbitrariness', *Journal of Social Philosophy* 48, no. 2 (2017): 197–211.

and Antarctica unsettles this. When humans govern over a territory with no humans, yet brimming with life, how should we think of the justification for exercising such a right? How should we think of the duties that the territorial agent should fulfil? And what should a governance arrangement that fulfils these rights and duties look like in practice? In Antarctica, nature should not be an afterthought, but a main character in territorial governance. And this should also be the case beyond Antarctica.

1.3 Some words on method and terminology

While grounded in political philosophy, this book draws from a range of disciplines. To offer a systematic picture of the emergence of Antarctica as a political territory, it is inevitable to visit the different phases of its history: seal hunting grounds and whale-hunting waters; *terra incognita* for avid European explorers; *terra nullius* for states eager to control new territories; a continent devoted to peace and scientific cooperation, and—arguably—the last unspoiled wilderness on earth. To clarify the structure and functioning of the Antarctic Treaty System, I look at its legal corpus and at the documents leading to its development. To engage in a critical assessment, I use analytical tools from international relations, political science, and geography. Finally, I draw from animal and environmental ethics and political theory to suggest what it would look like to have a political order that aimed for the representation of nature in Antarctica, and hopefully beyond.

I understand political philosophy as a sub-discipline within moral philosophy concerned with the conceptual, evaluative, and normative aspects of politics. What is justice, what counts as legitimate authority, who should have rights over a given territory and why? These are the kinds of questions that engage political philosophers. While I agree that the scope and normative ambition of political philosophy differ from those of moral philosophy, I thus see them in continuity rather than as radically different methodological enterprises.[19]

[19] On this I depart from theorists who agree that the substantive domain of moral and political philosophy is hard to distinguish in any principled way, but nonetheless believe that they are pulled apart by their *conditions of theorizing* (Christian List and Laura Valentini, 'The Methodology of Political Theory', in *Oxford Handbook of Philosophical Methodology*, ed. Herman Cappelen, Tamar Szabó Gendler, and John Hawthorne, 525–50 (Oxford University Press, 2016), 527, https://doi.org/10.1093/oxfordhb/9780199668779.013.10). See also Bernard Williams, *In the Beginning Was the Deed: Realism and Moralism in Political Argument* (Princeton University Press, 2005). Although this is not the place to defend one view over the other, I think the claim that political philosophy is constrained in a way that

This takes me to a second clarificatory point, regarding the level at which the theorizing is done along the spectrum between ideal and non-ideal; namely, to what extent the theory should be tied to real-world constraints, and to what extent it should be applicable here and now.[20] In this book, I keep John Rawls' idea of a 'realistic utopia' as a guiding principle when thinking about both Antarctic governance and global governance more generally.[21] Political philosophy must consider the physical and psychological limitations of humans and, to that extent, must be grounded in empirical evidence. At the same time, theories should not shy away from showing us moral signposts that, even if unattainable under current circumstances, are not beyond possibility. In other words, 'soft feasibility constraints' (related to societal, political, and economic circumstances) should not deter us from being ambitious in our moral and political theorizing.[22] If some desirable goal seems impossible to achieve, this may be a sign that we should use our critical and creative political imagination and acknowledge our duties to facilitate that transition, rather than settling for the status quo or something close to it.[23]

There is another level at which I 'idealize' that should be mentioned here and should be kept in mind in the coming chapters. Although I am not writing the Antarctic Treaty's hagiography, those critical to its structure and functioning may get that impression. The view that I present may seem extremely naïve, ignoring the ruthless geopolitical realities surrounding it, and ignoring that the Treaty's performance is far from optimal on many fronts and has moved extremely slowly in the last decade or two. Going back to the quote above, Antarctic governance has been created and sustained by real-world politics, as many have already persuasively shown.[24] I hope it becomes clear as the book progresses, however, that I do not deny this, but stand in amazement at its

moral philosophy is not is somewhat overstated. Instead, their relationship should be that of a *minima moralia* within a *maxima moralia.*

[20] Laura Valentini, 'Ideal vs. Non-Ideal Theory: A Conceptual Map', *Philosophy Compass* 7, no. 9 (2012): 654–64, https://doi.org/10.1111/j.1747-9991.2012.00500.x.

[21] John Rawls, *The Law of Peoples* (Harvard University Press, 1999), 128.

[22] Pablo Gilabert, 'Justice and Feasibility: A Dynamic Approach', in *Political Utopias: Contemporary Debates*, ed. Kevin Vallier and Michael Weber, 95–126 (Oxford University Press, 2017), 95.

[23] On critical and creative political imagination (of which I say more in chapter 9), see Avshalom M. Schwartz, 'Political Imagination and Its Limits', *Synthese* 199, no. 1 (2021): 3325–43, https://doi.org/10.1007/s11229-020-02936-1.

[24] See, for example, Klaus Dodds, *Pink Ice: Britain and the South Atlantic Empire* (I. B. Tauris, 2002); Adrian Howkins, *Frozen Empires: An Environmental History of the Antarctic Peninsula* (Oxford University Press, 2016); Aant Elzinga, 'The Continent for Science', in *Handbook on the Politics of Antarctica*, ed. Klaus Dodds, Alan Hemmings, and Peder Roberts, 103–24 (Edward Elgar Publishing, 2017); and Shirley V. Scott, 'Ingenious and Innocuous? Article IV of the Antarctic Treaty as Imperialism', *The Polar Journal* 1, no. 1 (2011): 51–62.

results. In other words, I celebrate the Antarctic Treaty as an institutional feat that has been achieved *despite of*, rather than *because of*, the actors involved. And yet, because these actors stand proud behind those accomplishments, I take their word for it and entice them to be even more daring with the interpretation of those legal documents that they are already committed to.

Throughout the book, I am especially mindful of two common biases in political philosophy, as well as in the actual performance of politics and law. I call these biases *methodological* because they inform the questions asked, they limit the units of analysis under consideration, and they constrain the answers given and their normative implications. The first is methodological nationalism, namely, a cognitive bias whereby the nation-state is treated as the natural unit of social and political organization, leading to the reification of state-centred conceptions of territory, space, and community.[25] As will be seen in the coming chapters, methodological nationalism is a constant feature in the way in which Antarctic politics and the theorizing around it have been conducted. This is no surprise and no exception in a world almost entirely divided into nation-states, but it is worth thinking about how it limits political possibility and imagination. It is telling that the most daring and sustained critique of the Antarctic Treaty, as shown in Chapter 4, was about handing power over Antarctica from the Antarctic Treaty members to the United Nations at large. What was questioned, thus, was not *that* states ought to be ruling over it, but rather *which* states. I spell out in more detail throughout the book why this way of thinking is problematic. Shortly stated, methodological nationalism prevents us from conceiving of alternative, more adequate, political institutions for responding to some of the greatest environmental challenges that we are

[25] Andreas Wimmer and Nina Glick Schiller, 'Methodological Nationalism and Beyond: Nation-State Building, Migration and the Social Sciences', *Global Networks* 2, no. 4 (2002): 301–34; Alex Sager, 'Political Philosophy Beyond Methodological Nationalism', *Philosophy Compass* 16, no. 2 (2021): e12726, https://doi.org/10.1111/phc3.12726; Alex Sager, 'Methodological Nationalism, Migration and Political Theory', *Political Studies* 64, no. 1 (2014): 42–59, https://doi.org/10.1111/1467-9248.12167. Some may object to my use of methodological nationalism as a term that has become fashionable but does not have much analytic purchase. They may say that, if the claim is empirical (namely, that states are in fact the most powerful actors in today's world), then few would deny it. Instead, if the claim is normative (namely, that states should be the only legitimate form of political order), then few would sustain it. I think that this objection, however, misses the target of the critique, which is that methodological nationalism makes us *think* in terms of states, *formulate our questions* around states, and *give answers* that assume that states (taken individually or as a system) are the default form of territorial governance. Even if we think of more independent international organizations, it is states that compose them: they are, after all, inter-*national*. As Alex Sager puts it, 'awareness of methodological nationalism's biases allows social scientists to scrutinize their axioms and findings. States continue to matter, but not necessarily in the ways that they contend' (Sager, 'Political Philosophy', 3). I thank an anonymous referee for pressing me to clarify this point.

facing. Its limitations become glaringly obvious when imposed over a permanently uninhabited and (mostly) uninhabitable place like Antarctica.[26]

The second bias is methodological anthropocentrism. Anthropocentrism has been extensively criticized in environmental ethics, environmental political theory, and animal ethics, but less so in standard political philosophy, politics, and law, where everything that is natural and non-human is almost without exception treated as a natural 'resource'.[27] Anthropocentrism refers to the belief that humans are the most important (if not exclusive) entities worthy of moral consideration, which leads in turn to a normative framework in which humans reign supreme. Correspondingly, I understand methodological anthropocentrism as the cognitive bias that turns humans into the main (and often only) units of analysis, leading to the neglect of other categories of living and nonliving natural entities. Why this is problematic will become clearer as the book progresses, but one obvious first answer is that this way of conceiving of and regulating our relationship with nature is inadequate. If the purpose of creating territorial rules is to keep peace, establish justice, and let those subject to them live better lives, then clearly there is something in this approach that needs revision.[28]

[26] I call Antarctica the only 'permanently uninhabited and (mostly) uninhabitable' continent on earth to mark two facts: first, what distinguishes Antarctica is not that there are no humans (in winter the population is around 1,000 and in summer it peaks at around 3,000), but that no permanent group of humans has stayed there over generations. Second, only a minimal part of the continent lends itself to human habitation, and even those parts are extremely challenging for human survival.

[27] Systematic critiques of anthropocentrism in environmental ethics are found in Richard Routley, 'Is There a Need for a New, an Environmental Ethic?', in *Reflecting on Nature: Readings in Environmental Ethics and Philosophy*, ed. Lori Gruen, Dale Jamieson, and Christopher Schlottmann (Oxford University Press, 2013); Val Plumwood, *Feminism and the Mastery of Nature* (Routledge, 1993); J. Baird Callicott, *Thinking Like a Planet: The Land Ethic and the Earth Ethic* (Oxford University Press, 2013); and Holmes Rolston III, *A New Environmental Ethics: The Next Millennium for Life on Earth* (Routledge, 2012). For environmental political theory aspiring to move beyond the limits of anthropocentrism, see Robert E. Goodin, 'Enfranchising the Earth, and Its Alternatives', *Political Studies* 44, no. 5 (1996): 835–49, https://doi.org/10.1111/j.1467-9248.1996.tb00337.x; Robyn Eckersley, *The Green State: Rethinking Democracy and Sovereignty* (MIT Press, 2004); and John S. Dryzek, *The Politics of the Earth: Environmental Discourses* (Oxford University Press, 2022). For two classics in animal ethics that are critical of the anthropocentric stance, see Peter Singer, *Animal Liberation* (Harper Collins, 2002) and Tom Regan, *The Case for Animal Rights* (University of California Press, 2004).

[28] Methodological nationalism can be pulled apart from nationalism as a political ideology, just like methodological anthropocentrism can be pulled apart from anthropocentrism as a moral stance. Regarding the former, for example, one can advocate cosmopolitan principles while holding methodological nationalist assumptions unquestioned. Inversely, one can be mindful of methodological nationalism and conclude nonetheless that the national community ought to be the preferred way of political organization (Sager, 'Political Philosophy'). Regarding methodological anthropocentrism, one can be an animal rights defender who starts from methodological anthropocentric assumptions; for example, by giving moral consideration only to non-human animals that share a specific human trait, like rationality. Inversely, one can be aware of methodological anthropocentrism as a cognitive bias while still concluding that moral theory should remain anchored in anthropocentric principles.

Regarding words, I have made some choices that may seem unusual but that I think necessary to strengthen the points this book wishes to make. Although some prefer the terms *other-than-human world* or *more-than-human world*, I opt for *nature*: the other two convey, in my view, more than merely natural entities, and connect to a trend of new materialism that assigns agency too profligately. Instead, one conclusion of this book will be that, for better or worse, it is (some) humans and (some) humans only who have the normative power to act and craft the rules of interaction between humans, and between humans and others.[29] Acknowledging that we have this particular kind of agency should not make us less, but rather more, concerned with those whose lives are relevantly affected by our decisions. Recall Spiderman: 'with great power comes great responsibility'.

When it comes to organized political collectives, I avoid using the language of *nations* and *nationalities,* insofar as this is possible. Despite growing trends in the opposite direction, I believe these categories are more harmful than helpful. This is not the place to expound on these ideas, but one quick defence of this decision is that it is one way to be mindful about, and to avoid falling into, uncritical methodological nationalism. When I can, thus, I describe the relations between states as *intergovernmental* rather than as *international*, and I refer to *state* rather than to *national* delegates or representatives. Admittedly, *international law* retains its name, although I think it should eventually be rebranded as *global* or *cosmopolitan law*; and the *United Nations* remains the United Nations, although the *United States of the World* would be a much better name.

Finally, one dangerous feature of adopting methodological anthropocentrism is that it tends to turn everything that is not human automatically into something for human use. This is why I avoid using the language of *natural resources*, and I often use scare quotes to remind the readers that those 'resources' are oftentimes beings with lives of their own.

1.4 Disclosure

I am not impartial when it comes to Antarctica. Born and raised in Punta Arenas, the Chilean 'gateway city' to the continent, as a child I was surrounded

[29] On this point, I am sympathetic to Andreas Malm's critique of some 'isms' in the humanities. See especially chapter 3 in Andreas Malm, *The Progress of This Storm: Nature and Society in a Warming World* (Verso Books, 2018).

by news about Antarctica. I knew people who travelled to Antarctica (most of them connected to the Chilean Air Force and Navy). I watched the daily weather report which included Villa Las Estrellas along with the other towns in the region, officially called 'Magallanes y Antártica Chilena'. And I studied the official map of the country, where Antarctica occupies most of the third column (Chile is so long that for it to fit onto any normal-sized poster, it needs to be cut in three).

I will never forget when, at school, we greeted the last huskies to go to the continent on an expedition, before the Environmental Protocol banned them along with all non-native species (from which *Homo Sapiens* was conveniently excepted). And I will never forget how puzzled I was when my father brought me a children's magazine from Argentina showing *their* Antarctic territory, almost identical to *ours*.[30] Never was it mentioned, during twelve years of schooling, that Antarctica was a territory ruled by many states, under the Antarctic Treaty. 'Chilean' Antarctica occupied my whole childhood imaginary.

Later, working as a journalist, I had the opportunity to visit Antarctica and report on life in Villa Las Estrellas, the Chilean settlement founded in 1984 on King George Island and inhabited by a rotating population of families from members of the national Air Force. It was the year 2000 and a school was still functioning. I went to a christening in the local church, a cosmopolitan football match in the gym, a tea party hosted by the housewives, and I even paid a short visit to the Russian base, just a few metres away from the Chilean one. There, I was greeted with vodka and taken on a tour of an underground cinematheque of Soviet films. I also joined the Chilean base's leader on a helicopter tour over Deception Island and the penguin-spotted icebergs along the coast. From above, I saw the eerie remains of the whaling stations and the hot springs where tourists used to bathe.

More than a decade passed before I went back to Antarctica, metaphorically, through my first big research project, 'Political Philosophy Looks to Antarctica'.[31] As a philosopher, it has always intrigued me how people make property claims and territorial claims over nature and how we seek to justify them. It was at some point during those investigations that I thought of Antarctica as a testing ground for normative theories of territory. Soon enough, I realized that the unique history of the continent and the way in which it

[30] Argentina, Chile, and the United Kingdom, I would learn much later, have overlapping claims over the Antarctic Peninsula and its surroundings.

[31] 'Political Philosophy Looks to Antarctica', University of Oslo, last modified 13 June 2022, https://www.hf.uio.no/ifikk/english/research/projects/political-philosophy-looks-to-antarctica/.

entered human politics were not just fit for testing the idea of territory but also related ones like colonialism, political representation, and environmental protection. Embarking in a second project, 'Dynamic Territory', I decided to use Antarctica to think systematically about some of the most pressing problems that we are facing as a global society; namely, how to think of territorial governance in a time when geography, climate, and demographics are shifting, many times in unforeseeable ways.[32]

I should also add that in the last three years, I have been participating in a global civil society initiative to draft a Declaration for the Rights of Antarctica.[33] As a philosopher, I am still struggling to marry the concerns of animal rights theories together with more ecocentric approaches in a harmonious way. However, despite my doubts, I still think there is political value in standing behind this cause. I am incredibly grateful to the members of the group for constant discussion of these topics, but I must underline that what I present here are my own views, not necessarily shared by all of them.

Finally, I am not sure that I would have written this book and got as excited about its subject matter had I not grown up with the Antarctic winds chilling my face. To the extent that territorial attachments may be formed remotely, I thus consider myself an Antarctican, and hope that at least some of the readers of this book will become Antarcticans too.

1.5 Structure of the book

The book has three parts. In Part I, *Occupation*, I look briefly at the history of Antarctica and examine the grounds given for territorial claims in Antarctica by seven states during the first half of the twentieth century. I propose that these claims can only be understood within the framework of late global colonialism. I then show how, unable to mutually acknowledge them, the claimant states plus a few others ended up signing a treaty that froze the possibility of individual sovereignty, but also that of internationalization under a Common Heritage of Mankind regime. This effectively turned Antarctic territorial governance into a one-of-a-kind.

In Chapter 2, I offer a historic overview of how Antarctica became a 'resource frontier' for seal hunters and whalers just over 200 years ago. I show

[32] 'DynamiTE: Dynamic Territory', University of Oslo, last modified 4 September 2025, https://www.hf.uio.no/ifikk/english/research/projects/dynamic-territory/index.html.

[33] See Antarctic Rights, 'A Voice for Antarctica', 2025, https://antarcticrights.org/.

how, during the first half of the twentieth century, a mix of political, economic, and scientific interests led seven countries to make territorial claims over almost the entire continent. I examine the two types of arguments given to justify those claims. On the one hand, there were connection-based arguments purporting to show some morally relevant link between the claimant and the land being claimed. These were concrete actions undertaken on the continent (for example, discovery and first occupation, the establishment of small settlements, etc.), but also legal documents claiming the existence of previous rights over those lands. On the other hand, there were function-based arguments that appealed to the activities and presence of the claimant states in their claimed areas—for instance, handling licences for economic exploitation and maintaining meteorological stations. I show that, even in the most charitable reading, these arguments are not enough to justify the extent of the claimed slices: an obvious justificatory gap remains.

In Chapter 3, I suggest that the only way to make sense of territorial claims in Antarctica is within the framework of late global colonialism, whereby the last unclaimed bits of land on earth were brought under the jurisdiction of individual states. In Antarctica, this hoarding of land was formally justified through the geographical doctrines of continuity, contiguity, and the sector principle—the same that were used by colonial powers elsewhere. While the wrongs of colonialism have been standardly associated with the subjection of one group of people by another, in Antarctica there were no people. So, was Antarctic colonialism wrong? I suggest that it was. While humans and non-humans have always been colonizing the earth, colonialism happens when there is an intentional and institutionally facilitated hoarding of land and resources that cannot be reasonably justified to others. As an attitude or disposition of character, colonialism happens when the *need* to occupy space and to take, use, and secure resources turns into *greed*. Without doubt, the most egregious shape that colonialism takes is when humans themselves are treated as resources for other humans, but this does not exhaust its wrongs. Towards the end of the chapter, I show how the inability of the claimants to agree among themselves on the legitimacy of their respective claims led to growing tensions in the 1940s and 1950s and, eventually, to the negotiations at the origin of the Antarctic Treaty.

In the second part of the book, *Representation*, I show how the Antarctic political community was constituted by virtue of the Treaty with a few countries acting as representatives of humanity as a whole. In moving forward, I suggest that Antarctic sentient beings and Antarctica as such should also be represented in the Treaty and other international fora where the decisions made may relevantly affect them. There would be important political, legal, and institutional challenges to surmount. Yet, the promise would be the creation of a political system that goes beyond the methodological nationalism and anthropocentrism that have so far been the default.

In Chapter 4, I present the way in which the political community and its representatives were constituted for Antarctica, through the Antarctic Treaty. By virtue of it, a group of countries appointed themselves as representatives of humanity for the continent, on the implicit condition that they would not derive any immediate economic gain from it. Instead, they would conduct science and keep peace for the benefit of all. I then examine the main criticisms of the Treaty that arose when the idea of negotiating a Mineral Convention began to be discussed and the prospect of economic gain appeared. Accusing the Antarctic Treaty of being an elitist club, a group of countries excluded from the arrangement demanded representation or, at least, a fair share of the benefits. I suggest that while these criticisms highlighted important issues of global inequality inherited from the colonial era, they did not question methodological nationalism and methodological anthropocentrism. The Environmental Protocol that ultimately replaced the failed Mineral Convention, I conclude, provides some of the tools needed to move beyond these methodological biases, even though so far these tools have not been used for those purposes.

Starting from one of the most discussed lines in the Environmental Protocol (namely, the claim that Antarctica has intrinsic value), in Chapter 5 I consider the possibility that the Antarctic political community ought not only to include humans but also to non-human sentient animals. I examine how these new members may be represented, either by having their interests internalized by the already existing delegates or by proxy representation. Favouring the latter option, I present a challenge for anyone attempting to incorporate non-human animals into the political community; namely, that depending on how we define their relevant interests, the prescriptions obtained will not just be different, but contradictory. This suggests a system of representation that might need to be more complex and diverse than what those endorsing the idea of representing non-human animals have granted so far. I conclude by suggesting that limiting the Antarctic demos to sentient beings only might not be enough, and that we should instead look at Antarctica itself as a natural entity worthy of political representation.

In line with a global movement advocating the rights of nature, in Chapter 6 I examine the claim that Antarctica—as one thriving community of life—should be considered as the subject of political and legal representation. I present some defences for the rights of natural entities in domestic constituencies and point to important differences between them and the Antarctic case. I consider the political, legal, and institutional challenges of considering Antarctica as a political subject, but also its promise: to overcome methodological anthropocentrism and methodological nationalism, and to create a precedent for other natural entities and places being granted political and legal recognition at the global level.

Part III, *Protection*, deals with the question of how effective (or not) the Antarctic Treaty and its associated instruments have been in keeping Antarctica as the purportedly most protected territory on earth. On the one hand, I suggest that the Antarctic model of Joint Guardianship geared towards protection rather than exploitation of nature should be exported to other places, most obviously ecoregions. These are sites that are particularly important for their biodiversity, their capacity for carbon storage, and the provision of other ecosystem services.[34] On the other hand, I point to the 'paradox of protection', whereby the continent, despite its tight environmental regulations, is dramatically affected by climate change even to the point of becoming a threat to others (by the rising sea-levels triggered by the melting of its ice). This suggests the need to stop seeing Antarctic governance as disconnected from the rest of the world, and to realize that protecting Antarctica requires acting beyond it.

Chapter 7 examines the claim that Antarctica is the most protected territory on earth thanks to the Environmental Protocol. I suggest that, without intending it, the parties consolidated a unique governance structure: first, they kept Antarctica as a territory beyond any individual state's jurisdiction, and agreed there would be no exploitation of mineral resources so long as it was in force. They opted, in short, for guardianship geared towards protection rather than sovereignty geared towards exploitation. I then go on to suggest that this model may serve as inspiration in places beyond Antarctica, particularly, in ecoregions like tropical rainforests. I address four objections to Joint Guardianship—namely, that such a solution would be unfair to the states in which these ecoregions lie; that it would be unfair to the local populations, potentially subjecting them to a new wave of colonialism; that it is politically unfeasible; and that there may be other models better suited than the Antarctic one. I conclude by reviewing the merits of the model while pointing to its limits and, in light of the discussion in the previous chapters, to the need to consider nature as worthy of representation.

Chapter 8 examines the 'paradox of protection' in Antarctica, namely, how the (arguably) most protected environment in the world is, notwithstanding, also one of the most threatened because of climate change, for which the very

[34] Dinerstein et al., 'Global Safety Net', 1.

signatories of the Environmental Protocol are largely responsible. While the Environmental Protocol has been hailed as the most ambitious legal instrument of its kind, I argue that its goals are impossible to achieve given its scope of application—south of 60 degrees south. To be effective, the Protocol must be interpreted in a more ambitious manner to eliminate its current 'problem of fit' (arising because of the incongruity between the way in which the ecosystem functions and the characteristics of the institution designed to protect it).[35] Specifically, I propose that, insofar as anthropogenic climate change is one of the largest threats to Antarctica, parties should reduce their greenhouse gas emissions not just *in* Antarctica but most importantly *at home*. I suggest that Antarctica may be said to have 'a right to be cold', in the sense that its living communities have a right to be protected from disruptive, avoidable change. This issues in a human duty to take measures based on the best available knowledge to prevent anthropogenic climate change. Although our best efforts at shielding Antarctica may not be enough, we ought to do what we can: this is partly what it means to protect. I conclude that, while the Environmental Protocol has been relatively successful at constraining damaging human action *in situ*, fulfilling its goal and purpose requires *ex situ* action. There are many other places around the world that are subject today to the same paradox.

Antarctica has been imagined as everything from the last treasure trove of resources for humanity to a green metropolis for sustainable growth to a sanatorium for the treatment of respiratory conditions to a nuclear waste dump. In Chapter 8, I take seriously the insights of those who have seen in the Antarctic model a promise, an open door to reconceive territorial governance along different lines than those that have been imposed over the rest of the world. Recapitulating the main arguments of the book, I conclude that Antarctic values should illuminate what a more dynamic global territorial order could look like, with states and other territorial entities collaborating much more tightly, and with the emphasis put on guardianship geared towards protection rather than on sovereignty geared towards exploitation. The uniquely creative Antarctic political imagination, furthermore, should help to move both Antarctic governance and global governance beyond methodological anthropocentrism and methodological nationalism, giving nature a representation that has long been due.

[35] I take the term 'problem of fit' from Oran R. Young, 'Environmental Governance: The Role of Institutions in Causing and Confronting Environmental Problems', *International Environmental Agreements* 3 (2003): 377–93, 378.

2
Claiming Empty Space

The first man to set foot on the Antarctic mainland was arguably Alexander Von Tunzelmann, a 17-year-old sailor from New Zealand. The episode is mentioned in polar history books as one-of-a-kind, given the peculiar circumstances under which it happened.[1] Von Tunzelmann came to Antarctica in January 1895 on board the Norwegian-owned *Antarctic*, a whaling ship that managed to cross the sea ice and reach Cape Adare, north of Victoria Land. After anchoring the ship, seven men took a longboat to get to the shore. Two of them were especially eager to be the first to jump: the Norwegians Leonard Kristensen, captain of the ship, and Carsten Borchgrevink, the geographer and explorer. So passionate was the debate between them that the young New Zealander sailor had to jump first to prevent the boat from capsizing. Although von Tunzelmann lacked the *animus occupandi* that inflamed Kristensen and Borchgrevink (who never recognized him as having been the first on shore), he became nonetheless a name to remember in the annals of Antarctic expeditions: one more in a long list of 'firsts', together with others like James Cook, Nathaniel Palmer, Roald Amundsen, and Richard E. Byrd.[2]

Being 'first' in Antarctica was not just a matter of individual honour or prestige; it was also and above all a political act that—depending on the conditions under which it took place—could entitle one's country to claim vast swaths of territory. The United Kingdom thus made the first territorial claim to the continent based on discovery and first occupation. A first Letters Patent was issued in 1908 (although it had to be amended in 1917, having initially included a slice of Chilean and Argentinian Tierra del Fuego).[3] It was also based on discovery and first occupation that France and Norway followed, in 1923 and 1938. The claims made by Chile and Argentina over Antarctica, in 1940, were seemingly different, but are ultimately also reducible to being 'first': not in discovering, or

[1] Olivier Remaud, *Thinking Like an Iceberg* (Polity Press, 2022), 89.

[2] Ibid. James Cook was the captain of the first European ship to cross the Antarctic circle (1775); Nathaniel Palmer was arguably the first to sight Antarctic mainland (1820); Roald Amundsen was the first to reach the South Pole (1911); and Richard Byrd was the first to fly over the South Pole (1929).

[3] J. Gordon Hayes, *Antarctica: A Treatise on the Southern Continent* (Richards Press, 1928), 364.

From Sovereignty to Guardianship. Alejandra Mancilla, Oxford University Press. © Alejandra Mancilla 2026.
DOI: 10.1093/9780197902769.003.0002

sighting, or landing, but rather in inheriting rights over the continent—rights first established by a fifteenth-century treaty between Portugal and Spain (of which they were ex-colonies). As members of the Commonwealth, finally, New Zealand and Australia were granted rights to Antarctica by the United Kingdom in 1923 and 1933 respectively, but they also felt the need to assert their own 'firsts' to legitimize the taking: Douglas Mawson's expeditions around the magnetic pole and long uncharted coasts, and von Tunzelmann's abrupt landing in Cape Adare, are profusely mentioned in their territorial narratives.[4]

Even today, proving to be 'first' in Antarctica remains a political issue. In a recent publication, Priscilla Wehi and colleagues have suggested that the real 'firsts' in Antarctica might have been the Māori.[5] The researchers refer to a sea voyage purportedly undertaken by the navigator Hui Te Rangiora as early as the seventh century AD. Traditional narratives, claim the authors, report that he and his crew met the 'frozen ocean', that they referred to flora and fauna unknown to their people but characteristic of the Southern Ocean, and that they made carvings of Tamarereti, protector of the Southern Ocean.[6] These claims have not been free of controversy, and some have responded denying that the Māori ever came close to Antarctica, and that 'at no stage was there indigenous occupation beyond about 50 degrees South ... it is most unlikely that Antarctic history began with pre-European Polynesian voyaging'.[7]

'So what?', one may ask. Why does it matter whether the Māori were there first or not? Why does it matter if it was Hui Te Rangiora rather than von Tunzelmann, or Borchgrevink, or Kristensen, who landed first on the Antarctic mainland?

The answer to these questions should become clear as this chapter progresses. I start with a brief history of how the Southern Ocean, the sub-Antarctic islands, and finally Antarctica entered Western maps from the eighteenth century onwards, thanks to expeditions like Cook's, but mainly because of seal hunters and whalers who kept sailing further south as they depleted the 'resources' further north. Since its discovery and systematic occupation 200 years ago, first by Western companies and later by Western powers, Antarctica became a testing ground for the doctrine of effective occupation (where being

[4] See, respectively, 'Australian Antarctic Territory', Australian Government (Geoscience Australia), last updated 30 November 2022, https://www.ga.gov.au/scientific-topics/national-location-information/dimensions/remote-offshore-territories/australian-antarctic-territory and 'First Landing on Antarctica', New Zealand History, updated 21 September 2021, https://nzhistory.govt.nz/media/photo/first-landing-antarctica.

[5] Wehi et al., 'Short Scan'.

[6] Ibid., 588 and 589.

[7] Anderson et al., 'On the Improbability', 3.

'first' was a key component). This doctrine legitimized the taking of land by colonial powers around the world during the second half of the nineteenth and the first half of the twentieth century, and it was stretched to the limit when applied to Antarctica—so much so that international legal scholars felt the need to develop a special understanding of polar occupation.[8] I then present two main types of arguments used to justify sovereign claims over Antarctica. On the one hand, there were those founded on the existence of some relevant connection with it—for example, through discovery, the performance of economic activities, and scientific exploration. While these connections were based on tangible actions, there were also connection-based arguments founded on legal documents invoking historical rights over the continent. On the other hand, there were function-based arguments that appealed to the claimant state's control and management of the territory—for example, through the granting of fishing licences and the administration of civil and criminal justice. I go on to show why not even the most charitable interpretation of connection and function-based arguments can justify the expanses of land claimed. I conclude that one can only make sense of them within a deeply colonial mindset.

2.1 From *terra australis incognita*, to hunting grounds, to claimed territory

Although the existence of Antarctica had been postulated since Antiquity, it was only speculation until 200 years ago.[9] Its discovery was in part due to scientific curiosity but more importantly due to economic drivers. The race towards Antarctic waters started among seal hunters who had overexploited the 'resources' around the Arctic and had to venture towards the other pole to find new hunting grounds. Sealers do not figure prominently in standard narratives of Antarctica, and it is hard to retell their story because (like whalers) they were secretive about their findings.[10] However, sealers and whalers are key to understanding how Antarctica entered the Western imaginary as a 'resource

[8] See, for example, W. Lakhtine, 'Rights to the Arctic', *American Journal of International Law* 24, no. 4 (1930): 703–17; Gustav Smedal, 'Acquisition of Sovereignty over Polar Areas', trans. Chr. Meyer, in *Skrifter om Svalbard og Ishavet no. 36* (I kommisjon hos Jacob Dybwad, 1931). While these authors refer to the Arctic as much as the Antarctic, the Antarctic was considered an especially hard case when it came to justifying effective occupation.

[9] Notably, around 340 BC Aristotle noted in his treatise on meteorology that if there was a north pole, then a south pole at the other end, where 'no one can live', presumably also existed: Aristotle, 'Meteorology', in *The Complete Works of Aristotle, Bollingen Series LXXI-2*, ed. Jonathan Barnes (Princeton University Press, 1984), 362a32–362b9, 587.

[10] Howkins, *Polar Regions*, 47.

frontier': an 'ongoing process of depletion and degradation of resources in one place followed by the search for new places to exploit . . . in which continual expansion is built into the business model'.[11]

Around 1786, British sealers arrived in South Georgia, shortly followed by the United States and France, and by 1810 the island was already a major centre for Southern Ocean sealing. While fur seals were sought for their pelts, elephant seals provided oil and blubber, needed for food production, lighting, lubrication of machinery, paint manufacture, and other industrial uses. There was no notion of sustainability or 'rational use' among the hunters, who acted as if the disappearance of animals in one site simply meant that they had moved somewhere else. Moreover, they had an incentive for short-term thinking, since they did not work for a fixed wage but for a share of the profits: the more they killed, the more they earned.[12] The seal industry, in short, had a 'rapaciously predatory character',[13] bringing the animals close to extinction.

Between 1775 and 1825, the gradual discovery of the Antarctic and sub-Antarctic regions was mostly due to seal hunters.[14] Not only South Georgia but also the South Shetlands, Deception Island, and the Antarctic Peninsula were all first sighted and discovered by sealers. In the words of historian Ben Maddison, '*acquisitiveness* rather than *inquisitiveness* generated the logic of geographical expansion that led to Antarctic exploration'.[15]

Whaling in Antarctic and sub-Antarctic waters, meanwhile, began in the late nineteenth century, with ships coming from Germany, Norway, the United Kingdom, Chile, and Argentina. From the turn of the twentieth century, whaling became by far the most important economic activity in the area, and as important for mapping the new continent as for scientific exploration. In 1904, funded by an Argentinian company, an onshore station was established by Norwegian captain Carl Anton Larsen in South Georgia, an island over which the British had taken nominal possession during Cook's visit in the late eighteenth century. By 1912, 'there were 6 onshore stations [in the Antarctic], 21 floating factories, and 62 whale catchers processing over 10,000 whales'.[16]

[11] Ibid. Howkins takes the idea of 'resource frontier' from Harold A. Innis, *The Fur Trade in Canada: An Introduction to Canadian Economic History*, rev. ed. (University of Toronto Press, 1970).

[12] Howkins, *Polar Regions*, 59–60.

[13] Ben Maddison, *Class and Colonialism in Antarctic Exploration, 1750–1920* (Pickering & Chatto, 2014), 36.

[14] Ibid., 101.

[15] Ibid., 10, his emphasis. See also María Jimena Cruz, Melisa A. Salerno, and Andrés Zarankin, 'Nineteenth-Century Connections Between Capitalism and Colonialism in Antarctica: The Case of Sealing in the South Shetlands', in *Colonialism and Antarctica: Attitudes, Logics, and Practices*, ed. Peder Roberts and Alejandra Mancilla, 195–224 (Manchester University Press, 2024), https://doi.org/10.7765/9781526170644.00015.

[16] Klaus Dodds, *The Antarctic: A Very Short Introduction* (Oxford University Press, 2012), 112–13.

Whale blubber fuelled the Industrial Revolution and became a key commodity, used to make candle wax and oil lamps, lubricants, and soap. During the First World War, whale oil was sought after to extract glycerine, a key ingredient in the manufacture of British artillery shells and small arms ammunition.[17]

It is no surprise, then, that what had remained as an open ground for the free taking of valuable resources became the last continent on earth where humans (or rather, states) would stake their claims. As Johan Nicolay Tønnessen and Arne Odd Johnsen note, 'whaling brought to a head the question of sovereignty in the Antarctic', completely changing its value.[18] Controlling Antarctica, especially its marine resources, became the goal.

2.2 Justifying territorial rights: function and connection

In a 2001 article entitled 'The Territorial Rights of States', A. John Simmons pointed to a problem that had gone almost unnoticed by political philosophers to date.[19] While political philosophy had been concerned with questions like 'On what grounds is political authority justifiable?', 'Should we obey the law?', and 'What is the best political system?', no one had cared to ask why a certain group of people ought to rule over a specific geographic domain in the first place. Territory, in short, was taken as a given that required no further justification. But this was not obvious, Simmons remarked, going on to propose his own individualist Lockean theory. In it, privately owned patches of land become a territory by the consent of their owners to come together under one political authority who, from then on, rules over them and their land.

Since then, the normative theorizing of territory has become a field of its own.[20] Central questions are what rights comprise territorial rights, what sorts of agents should be considered territorial agents, and what sorts of reasons are

[17] 'Why Whales Were Vital in the First World War', Imperial War Museums, 2025, https://www.iwm.org.uk/history/why-whales-were-vital-in-the-first-world-war.

[18] Johan Nicolay Tønnessen and Arne Odd Johnsen, *The History of Modern Whaling* (C. Hurst & Company and Australian National University Press, 1982), 178.

[19] A. John Simmons, 'On the Territorial Rights of States', *Philosophical Issues* 11 (2001): 300–26, https://doi.org/10.1111/0029-4624.35.s1.12.

[20] Some examples of normative theories of territory are David Miller, 'Territorial Rights: Concept and Justification', *Political Studies* 60, no. 2 (2012): 252–68, https://doi.org/10.1111/j.1467-9248.2011.00911.x; Cara Nine, *Global Justice and Territory* (Oxford University Press, 2012); Margaret Moore, *A Political Theory of Territory* (Oxford University Press, 2015); Anna Stilz, *Territorial Sovereignty: A Philosophical Exploration* (Oxford University Press, 2019); Avery Kolers, *Land, Conflict and Justice* (Cambridge University Press, 2009); Paulina Ochoa Espejo, 'On Borders: Territories, Legitimacy, and the Rights of Place' (Oxford University Press, 2020), https://doi.org/10.1093/oso/9780190074197.001.0001; Cara Nine, *Sharing Territories: Overlapping Self-Determination and Resource Rights* (Oxford University Press, 2022).

needed to justify holding territorial rights. In a nutshell, the standard answer to the first question is that the full package of territorial rights (characteristically held by sovereign states) includes rights of jurisdiction over people, rights to control borders, and rights over land and natural resources.[21] In terms of the territorial agents, these are normally thought to be politically organized collectives, paradigmatically states, but also sub-state groups like Indigenous communities. Finally, regarding the question of why these agents should enjoy those rights, two main types of arguments are given. On the one hand, as the name suggests, function-based arguments appeal to some important functions that the territorial agent performs and for which the agent needs some degree of control over a place. Guaranteeing the security of the population; giving access to essential goods like water, food, and a place to live; ensuring some level of well-being for the people; and establishing justice are all thought to be functions that merit granting territorial rights to those providing them. States are thought to be the main territorial agents able to fulfil these functions, and it is based on the performance of these functions that their 'full package' of territorial rights is justified. This way of justifying territorial rights has two features. First, although this is not conceptually necessary, function-based reasons tend to be invoked to justify the territorial rights of the territorial agents who are already in control of a place—more specifically, they are used to justify the territorial world order in its current shape. To put it in the terms introduced in Chapter 1, here methodological nationalism goes unquestioned. As Anna Stilz puts it in her justification of the territorial rights of states: 'The aim of the article is to make sense of an existing institution, the territorial state, which it presupposes.'[22] Given the world as we know it, the question then is: 'how do we now justify the authority of the state over the territory?' And (at least part of) the answer lies in looking at what the state is doing for its subjects. Second, *where* this happens (i.e. the territory itself) is taken for granted or—as its critics point out—left undetermined. Why is the territory *here* rather than *there* and why its borders are *these* rather than *those* are therefore questions that cannot be answered from a purely function-based perspective.

On the other hand, starting from this obvious justificatory gap, connection-based arguments appeal to some normatively relevant link between the

[21] This classification, which has become standard in the literature, is from Miller, 'Territorial Rights.'
[22] Anna Stilz, 'Why Do States Have Territorial Rights?', *International Theory* 1, no. 2 (2009): 185–213, 187, https://doi.org/10.1017/S1752971909000104. Although he does not frame his project in terms of the territorial rights of states and their justification, Allen Buchanan's work can also be read in this light: Allen Buchanan, *Justice, Legitimacy and Self-Determination: Moral Foundations for International Law* (Oxford University Press, 2004).

territorial agent and the place that justifies granting territorial rights to the agent *here*. The historical connection between the agent and the place, the labour that the agent has invested in the place, and the emotional attachment of the agent to the place are all reasons that, alone or combined, are invoked by this kind of argument. Connection-based arguments, then, match the agent with a specific geographical area: their whole point is to justify why the territorial rights of the agent lie *here* rather than somewhere else. Moreover, while states are obvious candidates to be territorial agents, they are not the only ones: connection-based arguments are particularly well suited to justify the territorial rights of individuals and non-state collectives, like Indigenous groups.

As could be expected, the most comprehensive and plausible justifications for territorial rights that have been developed so far are hybrid, in the sense that they appeal both to function-based and connection-based arguments. They thus aim to fulfil the 'particularity requirement'[23] and the 'generality requirement': respectively, they can answer why a certain agent holds territorial rights in a specific place by virtue of having normatively relevant connections, but also by virtue of performing some general functions that one should expect any territorial agent to perform.[24] I will say more about who the territorial agents of Antarctica have been and should be, and which specific rights they have held and should hold later in the book. For now, I examine the arguments offered by those states that claimed territory in Antarctica during the first half of the twentieth century. As will be seen, although the typology offered above was not available to them, they all implicitly or explicitly invoked relevant functions and connections to make their cases.

There are two important clarifications before proceeding. One is that many of these functions and connections happened *post factum*, after the official claims had been made. In other words, much of the 'effective' part of the occupation happened *after* rather than *before* the claims had been made. The other clarification is that the reasoning provided by each state to claim territory in Antarctica is not to be found in the original documents effectively claiming the territory, but in the scattered works of public officials and legal scholars looking for plausible justifications to appropriate the most implausible place. What follows is thus a reconstruction that does not mean to be exhaustive,

[23] Stilz, 'Why Do States Have Territorial Rights?', 188.
[24] See, for example, Stilz, *Territorial Sovereignty*.

but to capture what these scholars and public servants saw as the most solid grounds for their country's claims.[25]

2.3 Ceremonies of possession

In the fascinating book *Ceremonies of Possession*, historian Patricia Seed shows how the methods of appropriating space and claiming territory used by European colonial powers from the fifteenth to the seventeenth century were far from settled.[26] Succinctly summarized, 'Englishmen held that they acquired rights to the New World by physical objects, Frenchmen by gestures, Spaniards by speech, Portuguese by numbers, Dutch by description.'[27] When they arrived in America, the Spanish used to read the *Requerimiento* to the natives, a declaration closely resembling the Islamic *jihad* that had been applied earlier on the Iberic Peninsula, demanding obedience and giving the natives the chance to submit to their authority. Having done this, they proceeded to occupy the land. The Portuguese, instead, used their cartographic knowledge as an argument for possession: those first drawing the maps ought to have the right to appropriate the charted areas. The French performed elaborate ceremonies, and the Dutch relied on detailed descriptions as proof of their better claim. As for the English, settling, farming, and 'improving' the land were taken as signs of occupation, trumping any rights of previous dwellers who had not settled, farmed, or 'improved' the place in the English way. These disparate methodologies often led to conflict, with competing possessors talking past each other. As Seed summarizes, 'representatives of France, Spain, Portugal, the Netherlands, and England each shared the unmistakable conviction that their and only their position was right and that all the others were bull-headed wrong.'[28]

By the time Europeans and others started performing 'ceremonies of possession' in Antarctica, the question of how to appropriate humanly uninhabited and mostly uninhabitable space arose. It did not make much sense to read a *Requerimiento* inviting the penguins to surrender to the new owners, nor to try and grow potatoes in the barren lands.[29] At the same time, the main

[25] A thorough exposition can be found in Victor Prescott and Gillian Triggs, *International Frontiers and Borders* (Martinus Nijhoff Publishers, 2008), 384–91.

[26] Patricia Seed, *Ceremonies of Possession in Europe's Conquest of the New World 1492–1640* (Cambridge University Press, 1995).

[27] Ibid., 179.

[28] Ibid., 190.

[29] Note, however, that the existence of a garden was one of the proofs invoked by English diplomats when claiming the Falkland Islands against the Spanish. See ibid., 36.

complication of appropriating space anywhere else (namely, the presence of human natives) was absent here. This 'unsettled' the newcomers, who were used to associating possession with dispossession.[30]

In claiming King George Island, in the South Shetlands, on January 16, 1820, Edward Bransfield hoisted the British flag, took possession of the land in the name of HM King George IV, his heirs, and successors, and buried a bottle with British coins (a ceremony closely followed later by other British, like the seal hunter George Powell and the polar explorer James Clark Ross).[31] Twenty years after, on January 21, 1840, the French explorer Dumont d'Urville interpreted in his own way the British procedure of taking possession. In nearing Adélie Land, he reported,

> *I did send immediately one of our men to hoist the tricolour flag on these lands which before us had not been seen by any human being and on which no man's foot had been set. According to an old custom, carefully observed by Englishmen, we took in possession in the name of France the neighbouring coasts and the ice that prevented us from landing. Although the abuse of such taking in possession was the reason why it is often regarded as ridiculous and of no consequence, yet in this case at least we considered ourselves to be sufficiently entitled to support the old custom on behalf of our country as we did not deprive anybody of their possession and our titles were unquestionable. Consequently, we considered ourselves to be on French territory.*[32]

Dubouzet, the second officer of the *Zélée*, on the same expedition, confirmed that it had been '"a wholly pacific conquest", undertaken with "none of the abuse which has been born of such acts" in many places.'[33]

These methods of symbolic annexation survived all the way into the twentieth century, even among countries that ended up making no formal claims. In 1938–1939, Germany sent an expedition to the area soon to be claimed by Norway and dropped markers claiming the land for the Third Reich:

> *Ernst Hermann, the geographer, later reported that the party landed on the ice shelf and raised the swastika a few hundred metres from the edge. 'This is the outward sign that we Germans have trod this no-man's land and claimed it*

[30] Maddison, *Class and Colonialism*, 52.

[31] Quoted in Jacek Machowski, *The Status of Antarctica in the Light of International Law* (National Center for Scientific, Technical and Economic Information, 1977), 54.

[32] Quoted in ibid., 55.

[33] Quoted in Maddison, *Class and Colonialism*, 52.

> *for Greater Germany', wrote Hermann. 'The first German colony!', he added. A 'native' waddled forward to meet them and they greeted him with a 'Heil Hitler!' 'The penguin', Hermann reported, 'was not impressed'.*[34]

But definitely the most ambitious were US aerial explorers: adamant that his country should get a slice of the continent, Richard E. Byrd gave the name of his wife to the land he sighted from above in 1929, while Lincoln Ellsworth lay an unofficial claim on behalf of his country to an area he overflew

> *by dropping from his plane a copper receptacle 8.5 by 2.5 inches in size, with a letter of the following content: 'To all and sundry: having flown in a straight line . . . I drop this note with a flag of the United States of America and advance in so far as this act permits a claim to the area south of 72 degrees of latitude and 79 degrees of longitude East, which areas I declare to have explored on January 11, 1939 A.D. Signed: Lincoln Ellsworth.'*[35]

These convoluted ways of 'taking possession', however, were not enough to count as effective occupation. The doctrine, well established by then, required the claimant states to show not just their *animus occupandi* over a territory through the performance of these symbolic acts, but also their *corpus occupandi*; namely, their tangible presence and activity in the place. To justify possession of a territory, states thus needed to provide a mix of connection- and function-based arguments: the existence of settlements, the use of the land for economic purposes, the establishment of state institutions, and the enforcement of law all went to show that the state's sovereignty was justified. In the case of the polar regions—and especially of Antarctica—legal theorists quickly realized, however, that the standards applied elsewhere would have to be relaxed or adapted. As noted by the legal scholar Lakhtine, the application of the rules for terrestrial territory to polar territory 'must be recognized as absolutely irrational'.[36]

As mentioned before, the first state to claim Antarctic territory was the United Kingdom, through Letters Patent in 1908, amended in 1917.[37] The main motive was control over the Southern Ocean and its whaling industry. As

[34] Robert E. Wilson, 'National Interests and Claims in the Antarctic', *ARCTIC* 17, no. 1 (1964): 15–31, http://arctic.synergiesprairies.ca/arctic/index.php/arctic/article/view/3481.
[35] Machowski, *Status of Antarctica*, 56.
[36] Quoted in Oscar Pinochet de la Barra, *Chilean Sovereignty in Antarctica* (Editorial del Pacífico, 1954), 32.
[37] The first Letters Patent were done in such haste that not only did they get the limits wrong but also the basic principles of geography, claiming land to the *south* of a meridian: Hayes, *Antarctica*, 364.

they would clarify in 1955, when presenting their case against Argentina and Chile to the International Court of Justice, these letters 'confirm[ed] past acts of discovery and annexation carried out repeatedly by British nationals, since Antony de la Roche discovered South Georgia in 1675'.[38] The naming of places was to them a clear proof of their entitlement: 'The first discoveries of South Georgia, the South Sandwich Islands, the South Orkneys, the South Shetlands, and Graham Land were . . . all made by British nationals—a fact reflected in the names given to these territories by which they have been known, and have figured in maps and charts, ever since'.[39]

The territories were placed under the administration of the Government of the Colony of the Falkland Islands, and since then, the state attempted to show its presence in different ways. A Whaling Code established in 1906 was replaced by a Whaling Ordinance in 1908, the aim of which was to protect whale stocks from being depleted. To do so, the British granted licences to foreign companies (Norwegian, among others) and sent public servants to the whaling stations to ensure that the regulations were respected. Among the other displays of state activity that the United Kingdom invoked were the creation of a similar code for sealing in 1909, and the establishment of a post office in the whaling settlement of Grytviken in 1912. How many letters were effectively sent and received was not mentioned.

This unilateral method of annexing territory was followed by all the other claimants: New Zealand (1923), France (1924), Australia (1933), Norway (1939), Argentina (1940), and Chile (1940). Except for Norway (which left its south border undetermined), they all claimed for themselves full wedge-shaped slices of the continent, reaching all the way to the South Pole. Only the most inaccessible part of the continent, Marie Byrd Land, remained unclaimed (it was part of the US strategy to not claim any specific areas, but to leave the possibility of future claims open).

Strictly speaking, New Zealand and Australia did not make claims over the continent but had the administration of their slices 'transferred' to them by the British.[40] Together with France and Norway, what they had in common was the grounding of their claims in being first on the land, either as representatives of

[38] International Court of Justice, *Antarctica Cases (United Kingdom v. Argentina; United Kingdom v. Chile)*, 1956, 12, https://icj-cij.org/case/26.

[39] Ibid.

[40] See *Australian Antarctic Territory Acceptance Act (1933)*, Australian Government, Federal Register of Legislation, latest version 31 December 1973, https://www.legislation.gov.au/C1933A00008/latest/text; and *Ross Dependency Boundaries and Government Order in Council 1923 (SR 974)*, New Zealand Legislation, reprint 17 August 1923, https://www.legislation.govt.nz/regulation/imperial/1923/0974/latest/DLM1195.html.

their states, as scientists and explorers, or simply as seal hunting and whaling entrepreneurs, indirectly advancing the interests of the nation. Argentina and Chile, on the contrary, based their claims only partly on what the state and its citizens had done *in situ*.[41] They underlined, instead, that they were not claiming territory, but enacting their historical rights—rights they had purportedly inherited from the days when they were colonies of Spain.[42]

In the next section, I show that not even the most generous interpretations of connection-based and function-based arguments would grant rights over the actual slices claimed by these states. There is thus a major justificatory gap that remains to be filled. While today those claims remain suspended under the Antarctic Treaty, they are considered by many as the thorn in the side, the unresolved issue, the latent problem that will emerge sooner or later.[43] In this context, it is worth asking whether these claims should be given up entirely.

2.4 The limits of connection-based arguments

Standardly, arguments to justify territorial rights based on connection appeal to physical acts and to the physical presence of the agent in the place. When it comes to Antarctica, as mentioned before, being the first to set foot on land was thought to give an undeniable preliminary claim over it, although merely sighting it was generally deemed good enough. This is why first exploration and discovery figured as part of the arguments given by the British, French, Australians, and New Zealanders. For the British, their initial discoveries created an original 'root of title' over the place.[44] The Norwegians mentioned Roald Amundsen, the first to reach the South Pole in 1911, who, during his expedition, 'discovered large areas that were named and taken into possession on behalf of the King of Norway'.[45] The French based their claim on Dumont d'Urville's discovery of Adélie Land in 1840; the Australians mentioned the Australasian Antarctic Expedition led by Douglas Mawson (1911–1914), and the New Zealanders highlighted their Antarctic 'firsts', from Tuati (who

[41] Argentina does underline, however, the existence of a permanent meteorological station in the South Orkneys, functioning since 1904.

[42] Shirley V. Scott, 'Imperialism in Antarctica', in *The Palgrave Encyclopedia of Imperialism and Anti-Imperialism*, ed. Immanuel Nez and Zak Cope (Springer, 2021), https://doi.org/10.1007/978-3-319-91206-6_142-1.

[43] See, for example, Alan D. Hemmings, 'Security Beyond Claims', in *Antarctic Security in the Twenty-First Century: Legal and Policy Perspectives*, ed. Alan D. Hemmings, Donald R. Rothwell, and Karen N. Scott (Routledge, 2012).

[44] International Court of Justice, *Antarctica Cases*, 12.

[45] Magnus Hovind Rognhaug, *Norway in the Antarctic* (Norsk Polarinstitutt, 2014), 7.

sighted Antarctic mainland in 1840 while on a US-led voyage) to Alexander von Tunzelmann, arguably the first to set foot on the continent, as described at the beginning of this chapter.[46]

A second source of connection-based arguments in Antarctica referred to the exploitation of its marine animals. Argentinians and Chileans invoked their seal hunting activities as demonstrating their presence in the area. Alberto Daverede, an Argentinian diplomat, recalled how, as early as the beginning of the nineteenth century, Argentinian seal hunters frequented sub-Antarctic islands 'to exploit the valuable skins of those wretched creatures who fell by the hundreds of thousands under the mallets of avid sea merchants that hid with zeal the cold places where they carried out the slaughter'.[47] The Chilean diplomat and Antarctic scholar Oscar Pinochet de la Barra noted that when seals became extinct in Tierra del Fuego, hunters went further south, all the way to the South Shetlands. In his words, 'the exploitation of the natural resources of such regions constitutes a completely satisfactory juridical basis [for the Chilean claim]'.[48] As for whaling, Norway's claim to Queen Maud's Land was 'legitimized in large part' by the work done by the *Norvegia* expeditions, a series of scientific voyages financed by the whaling entrepreneur Lars Christensen.[49]

Finally, a unique kind of connection-based argument used in Antarctica appealed to the historical link between the state and the place, no matter how immaterial. As mentioned above, this is how Argentina and Chile partly supported their purported rights to the continent. According to the doctrine of *uti possidetis iuris* (used in Latin America to draw the boundaries of the newly independent republics following the colonial demarcations), what Spain had taken as hers in the past was now theirs by right. This was the case with all the lands in the 'New World' lying as far as the South Pole, as established by the Papal Bull *Intercaetera* (from 1493) and the Treaty of Tordesillas with Portugal (from 1494). The fact that these lands were mere speculation at the time of the signatures did not lessen their validity, according to both parties. For the Argentinian ambassador Pedro Radío, writing in 1948, the 'Decree issued by the Pope was universally recognized, in his role of supreme authority

[46] See, respectively, 'Australian Antarctic Territory', Australian Antarctic Program, last updated 15 April 2016, https://www.antarctica.gov.au/about-antarctica/australia-in-antarctica/australian-antarctic-territory/, and 'Antarctica and New Zealand', New Zealand History, updated 2 July 2024, https://nzhistory.govt.nz/politics/antarctica-and-nz/people.

[47] Alberto L. Daverede, 'Política y Actividades Antárticas de La República Argentina', in *Aula de Estudios Antárticos* (Madrid, 1987), 3. This and all other translations from Spanish hereafter are mine.

[48] Pinochet de la Barra, *Chilean Sovereignty*, 32.

[49] Rognhaug, *Norway in the Antarctic*, 7.

of Christendom'.[50] For the Chilean Antarctic scholar and ambassador José Berguño Barnes, writing in 2000, the Bull 'fully conformed to the public law of its time', while the Treaty 'incorporated the bull as positive international law between the two kingdoms'.[51] It was a legal act with universal juridical effects.

I suggest that there are three main challenges for these connection-based arguments. The first and most important is how to justify the extent of the claims, considering the limited connections to the lands claimed. The second is whether the fact that Antarctica was turned into a resource frontier should count in support of (and not against) those who made it so. The third concerns the weakness of the appeal to an intangible historical connection, and to how a relevant change of circumstances should affect the justice of the claims.

2.4.1 The mismatch between the extent of the actual connections and the extent of the claimed territories

One criticism of connection-based arguments justifying territorial rights is that their persuasiveness dilutes the further away one moves from the human-inhabited centre. Arguments that appeal to some relevant link between the agent and the land mostly rely, after all, on the actual presence of that agent in that land: the lesser the presence, consequently, the weaker the link.

To understand the magnitude of the claims made by these seven countries in Antarctica, it is worth looking at Table 2.1. Four of the seven claimants claim larger areas in Antarctica than their own: more than eight times in the case of Norway, seven times in the case of the United Kingdom, and more than one and a half times in the cases of New Zealand and Chile. There is one country (Australia) that claims over 40 per cent of the continent, and another one (Norway) that claims almost a fifth. The areas claimed vary in extension from roughly the size of Paraguay (France's claim) to somewhere between the size of India and Australia (Australia's claim). If one looks at the actual connections invoked by the Antarctic claimants, however, they happened in very specific geographic sites, like Ross Island and the South Shetlands, and along some coastal areas, but hardly inland (except for the passing of a few expeditions). The question, then, is how to justify the claiming of huge wedges of territory by virtue of simply having been present in a minuscule part of it.

[50] Pedro Radío, *Soberanía Argentina en la Antártida* (Universidad de Valladolid, 1948), 43.

[51] Jorge Berguño Barnes, 'Historia Intelectual del Tratado Antártico', *Boletín Antártico Chileno* 19, no. 1 (2000): 2–12.

Table 2.1 Territorial Claims in Antarctica (in Square Kilometres)

Claimant	Country's Area (in km²)[a]	Area Claimed in Antarctica (in km²)	Percentage of Antarctica Claimed[b]	Percentage of the Antarctic Claim in Relation to Own Area
Argentina	2,780,400	1,461,597	10.29	52.6
Australia	7,741,220	5,900,000	41.5	76.2
Chile	756,102	1,250,257	8.8	165.3
France	643,801	432,000	3.04	67.1
New Zealand	268,838	450,000	3.16	167.3
Norway	323,802	2,700,000	19.01	833.8
United Kingdom	243,610	1,709,400	12.03	701.7

Note: The table shows the size in square kilometres of the claimant states, and the size in square kilometres of the areas claimed in Antarctica. It then shows the percentage of the continent that they claim, and the percentage of their claims in relation to the state's areas.

[a] CIA, 'Country Comparisons—Area', *The World Factbook*, https://www.cia.gov/the-world-factbook/field/area/country-comparison/.

[b] The percentages are calculated based on Antarctica as having a total area of 14,200,000 km^2.

If there is one philosophical theory that is useful to interpret claims to Antarctica based on being first and being present, that is John Locke's theory of property rights. For Locke, humans have a natural right to property that precedes any human law or institution. By 'mixing' our labour with the land, we become owners and acquire rights over the external world.[52] In the Antarctic narratives, the hard toil of explorers, scientists, seal hunters, and whalers and the concrete vestiges of their labour-mixing in Antarctica appear time and again. This suggests that a plausible way to answer the question above may be to look at Locke's theory and see what it has to say about the setting of limits.

In his analysis of Locke's theory of rights, A. John Simmons tackles this problem.[53] If mixing one's labour with the land, as Locke suggests, is what

[52] John Locke, 'The Second Treatise of Government', in *Two Treatises of Government*, ed. Peter Laslett (Cambridge University Press, 1988), first published anonymously in 1689, §27, 287–88.

[53] Interpretations and critiques of Locke's theory are numerous, and I cannot do them justice here. I choose Simmons's analysis because I find it to be one of the most generous towards Locke. If even one of the most charitable readings of Locke shows that there are stringent limitations on land appropriation, then this will reinforce the point I wish to make, namely, that connection-based arguments that rely on physical acts and presence, and on the exploitation of resources, cannot be invoked to justify Antarctic territorial claims as they stand. Having said this, Locke's theory is a theory of property rights and not territorial rights. These are derivative on them and require the agreement of property owners to subject themselves to a common territorial authority.

grounds the right to property—and, derivatively, territory—then how do we know how far it can stretch? Simmons's answer is: not very far. There is a '*maximum possible* extent of the property labour creates',[54] determined by the nature of the activity:

> *We can take that which is necessary to our projects (and perhaps reasonable windfalls of those activities), but our property runs only to the boundaries of our implemented projects (and not just to whatever we might envision).*[55]

In Simmons's example, if we plant a corn field, we acquire property rights not only for the area occupied by the corn plants themselves, but also for the land between and around the plants, to the extent that this is necessary to carry out the activity (if we find gold nuggets while ploughing, these might be considered a reasonable windfall). Applied to the Antarctic case, claimants could argue that the purpose of their activity was ultimately to appropriate as many resources as possible then and for the foreseeable future, not just for the individuals directly involved but also for their states. However, as Simmons notes, for Locke the right to land and natural resources is limited by the right to individual self-government.[56] To prevent some from taking so much that others are prevented from exercising their right to self-government, two provisos are put in place. First, one may appropriate land and natural resources only to the extent that this does not harm others: 'For this labour being the unquestionable property of the labourer, no man but he can have a right to what that is once joined to, *at least where there is enough, and as good left in common for others*.'[57] This is the *fair appropriation* proviso. Second, it is only justifiable to own what one labours and continues to labour as part of one's projects. What is spoiled, wasted, or unused should remain in or returned to the common. In Locke's words: 'As much land as a man tills, plants, improves, cultivates, and can use the product of, so much is his property. Whatever is beyond this, is more than his share, and belongs to others.'[58] This is the *non-waste* proviso.

Seen in this light, Antarctic claims clearly violate the *non-waste* proviso, and depending on how one interprets it, the *fair appropriation* proviso.

[54] A. John Simmons, *The Lockean Theory of Rights* (Princeton University Press, 1992), 276, his emphasis.
[55] Ibid.
[56] Ibid.
[57] Locke, 'Second Treatise', §27, 288, my emphasis.
[58] Ibid., §31, 290.

Regarding *non-waste*, the seven claims exceed by far what the claiming agents were using or argued that they would need to use. It is one thing to claim territorial rights over the island where some nationals perform whaling activities; it is another to extend those territorial pretensions all the way to the end of the world. It is one thing to claim that the existence of a permanently inhabited meteorological station gives the state rights over the place where the station is located. It is quite a different thing to claim that the existence of that station gives one rights over an adjacent territory the size of Peru.[59] Even the most modest claim to territory in Antarctica, that of France, seems unjustifiably inflated if one uses *non-waste* as a demarcation criterion.

Regarding *fair appropriation*, when it comes to land, the claimants could say in their defence that, at the time, they were not violating any proviso: nobody's right to self-government was infringed upon by the unilateral claiming of Antarctica. They did not dispossess any human being, and they did not make the situation worse for anyone in any obvious manner. By claiming almost an entire continent for themselves, however, they unilaterally left out other potential claimants (most of the world, in fact). Put differently, they unilaterally decided for others what 'enough and as good' amounted to when it came to appropriating the last part of the earth that was not yet under human governance.

In sum, most of the connection-based arguments used by claimants in Antarctica appeal to physical acts and to very limited presence in the continent. Even if one interprets very loosely the idea that 'purposive activities' give grounds to rights over land (as suggested by Simmons's reading of Locke), those arguments do not suffice to justify the huge expanses of land claimed. On the contrary, they remind us of Robert Nozick's story of the man who spilled a can of tomato juice in the ocean, hoping to become its owner.[60] Invoking physical connection, in short, could only serve to ground claims that are much more modest in extent than the existing ones.

2.4.2 Exploiting animals to gain territorial rights

A challenge for claimants like Argentina and Chile, who base their claims partly on economic exploitation, is to explain why exploiting a 'resource', even

[59] The United Kingdom exemplifies the first case, while Argentina exemplifies the second one.
[60] Robert Nozick, *Anarchy, State and Utopia* (Basic Books, 1974), 174–75.

to depletion, should give one territorial rights over the area where the exploitation occurs. This seems like an inversion of logic: should not the burden of proof be on the exploiters to justify their continued control over the 'resources' and the places that the 'resources' inhabit?

This challenge could, of course, be objected to as being anachronistic. At the time when these arguments were made, claimants could be excused for being non-culpably ignorant: notions of environmental sustainability (let alone of animal rights) did not yet exist, and they had a reasonable expectation that the 'resources' were almost inexhaustible. But this was not the case. As already mentioned, one of the most avid exploiters of whale oil and blubber, the United Kingdom, had already put a Whaling Ordinance in place in 1906 and a Seal Ordinance in 1909, realizing that, if unchecked, excessive hunting would drive the animals to extinction or near extinction.[61] It was clear, especially after the depletion of seals and whales in northern waters, that animals did not just keep popping up somewhere else after entire colonies had been swept away. Even by the standards of their own time, then, invoking exploitation as a ground for appropriation was problematic.

If one takes Locke's provisos to analyse this argument, *fair appropriation* is clearly being violated. In pursuing their activities, those engaged in seal hunting and whaling were definitely not leaving 'enough and as good' for others to take. (Notice that, for the sake of argument, I am assuming that hunting seals and whales was a morally legitimate activity; that is, I am not questioning the methodological and substantive anthropocentrism at work here.) Regarding *non-waste*, sealers often used to kill many more animals than they were able to use. Rotting carcasses left on the beach because there was no space for them in the vessels or because there was no time to process them was not an uncommon sight. In sum, even at the time it was recognized that excessiveness was the mark of this activity, and that this was problematic, if not for the animals, for the very continuation of the business.

2.4.3 Tenuous historical connections and changed context

While the previous connections are tangible, physical ones (taking and occupying the territory, exploiting marine 'resources'), the third kind of

[61] International Court of Justice, *Antarctica Cases*, 16–17. To be sure, their motivation was not purely altruistic or driven by environmental considerations. The United Kingdom was interested in granting licences and charging vessels a fee to be allowed to hunt in 'its' waters.

connection, based on legal documents, is much more ethereal. Chile and Argentina apply this kind of argument, which requires going 500 years back in history.[62]

Oscar Pinochet de la Barra, one of Chile's most renowned Antarctic scholars, explains that, until 1492, by virtue of a series of papal bulls, Portugal had enjoyed a monopoly on all newly discovered lands and seas in the regions of the world unknown to Europeans. This changed, however, after Columbus. In 1493, Pope Alexander VI granted the Kings of Castile and Aragon the right to all lands discovered by Columbus and drew the line between Portuguese and Spanish dominions at 100 leagues (approximately 550 kilometres) west of the Cape Verde Islands and the Azores, from the Arctic to the Antarctic. After the king of Portugal complained, the Treaty of Tordesillas of the 7th of June of 1494 changed the line to run at 370 leagues (approximately 2,000 kilometres) west of the Cape Verde islands. Portugal was to govern all lands discovered or to be discovered east of that line, and Spain those to the west.[63]

Although Antarctica remained a matter of speculation during the centuries to follow, different official documents from the Spanish colonies (like the Captaincy General of Chile)[64] mention the Antarctic pole as the limit of their jurisdiction. When these colonies became independent in the nineteenth century, they inherited the borders from the time when they were administrative units under Spanish rule. From this it follows, according to Argentinian and Chilean scholars, that they inherited Spain's Antarctic claim. In contrast to the other claimants, they never considered Antarctica as *terra nullius*, but as theirs by historic right.

If one assumes, for the sake of argument, that claiming territory merely by virtue of past paperwork is acceptable, there are still two problems with this line of reasoning. First, although the Treaty of Tordesillas was generally respected by the two signing parties, no other European power, Catholic or Protestant, ever acknowledged it. Defending a sea free for all (and especially for the Dutch), Hugo Grotius denied the universal validity of the Papal bull, noting that the Pope might well act as arbitrator between Portugal and Spain, but that this 'appertaineth not to the rest of the nations'.[65] On the same topic, the French legal scholar, Paul Fauchille, pointed out centuries later that:

[62] I provide a more detailed analysis of their claims in Alejandra Mancilla, 'South American Claims in Antarctica: Colonial, Malgré Tout', *The Polar Journal* 12, no. 1 (2022): 22–41, https://doi.org/10.1080/2154896X.2022.2062558.

[63] Pinochet de la Barra, *Chilean Sovereignty*, 9–10.

[64] Ibid., 14–19.

[65] Hugo Grotius, *The Free Sea, With William Welwood's Critique and Grotius's Reply*, trans. Richard Hakluyt, ed. David Armitage (Liberty Fund, 2004), first published 1609, 16.

> *Catholic princes who had not been favoured with papal gifts were discontented. The King of France, Francis I, requested to be shown the will of Adam which deprived him of the right to acquire land in the New World.*[66]

Second, even if other states had accepted the Treaty when it was signed, and even if the use of *uti possidetis* made sense for lands not yet discovered at the time of the transfer, legal documents should be subject to amendment as the relevant circumstances change. Even if dividing the 'New World' between two European powers had been a reasonable arrangement in 1494, this was no longer the case during the first half of the twentieth century, when the Antarctic claims were made. This is not to say that the underlying moral principles change with time, but rather that their application must be sensitive to context.

2.5 The limits of function-based arguments

By the turn of the twentieth century, effective occupation had become the accepted principle for states to justify the annexation of new territories. Legal historian Martti Koskenniemi points out that legal textbooks described this as a major progress at the time:

> *They criticized the exaggerated importance given in the sixteenth and seventeenth centuries to discovery and symbolic annexation and emphasized the rational basis of the rule that required effective presence [of the state] as the condition of territorial right.*[67]

In line with this, while some sort of relevant link was generally the basis on which countries staked their Antarctic claims, the performance of specific functions by the state in the territory was seen as cementing those claims. It was usually the case that, after the claim had been made, the state proceeded to mark its presence through different means. Among these were the appointment of public authorities and military personnel, the sponsoring of scientific expeditions, the establishment of permanent meteorological stations, the organization of rescue missions and the maintenance of navigational aids, the

[66] Quoted in Smedal, *Acquisition of Sovereignty*, 14.

[67] Martti Koskenniemi, *The Gentle Civilizer of Nations: The Rise and Fall of International Law 1870–1960* (Cambridge University Press, 2001), 149.

granting of licences and permits for economic activities, the issuing of postage stamps, and the mapping and naming of the area.[68]

Thus, even if connection-based arguments seem insufficient to justify the extent of the actual claims, claimant states could reply that what made them rightfully sovereign in their Antarctic territories was precisely the performance of these functions. In putting its case against Argentina and Chile to the International Court of Justice, the United Kingdom did just that. In a section entitled 'Display of British Sovereignty in or in Regard to the Dependencies in the Period July 21, 1908–September 22, 1938', the United Kingdom listed all the activities that it had conducted in the area since the promulgation of the first Letters Patent. The Falkland Islands Ordinance (which covered laws regarding the administration of civil and criminal justice, marriage, testacy and intestacy, etc.) was made *applicable* to the new annexed territories (even if it was hardly if ever *applied*); the regulation of whaling and sealing was presented as a display of the United Kingdom's environmental stewardship over the area, through the granting of licences to Norwegian, Argentinian, and Chilean companies; the appointment of a resident magistrate and customs and police officers in South Georgia, and the establishment of government buildings and even a marine laboratory were underlined, as well as the scientific expeditions carried out by research vessels. In those places where little or no activity was shown, the excuse was their almost utter inaccessibility.[69]

But may function-based arguments be successfully invoked to justify the claims of individual states in Antarctica? I suggest that they may not, insofar as they beg the question of why it is them individually, rather than them together, or with others, or anyone else, or nobody at all, that should be in control. To make their unilateral claims in the only humanly uninhabited and mostly uninhabitable continent on earth, the individual claimants took for granted not only that the place should be brought under some sort of jurisdiction, but that this jurisdiction was going to be divided among individual states, just like in the rest of the world. However, it is one thing to argue that the establishment of justice gives rights to a state over a territory where there is a human population in need of it. It is quite another thing to argue that the establishment of justice

[68] Emilio J. Sahurie, *The International Law of Antarctica* (Martinus Nijhoff Publishers, 1992), 259–77.

[69] International Court of Justice, *Antarctica Cases*, 16–21. Although they did not use the language of "environmental stewardship" at the time, it was clearly the purpose: "One of the chief objects of the legislation was the conservation of stocks by regulating the number and tonnage of whaling vessels, the number of whaling licences, the number of whales to be taken by each licence-holder, by protecting whale calves and by other measures" (ibid., 16).

over a distant and humanly uninhabited territory gives a state rights over it. As philosopher Cara Nine puts it:

> *territory cannot be established in areas where there are no persons (as 'legitimate rule of law' implies provision of basic needs for persons), or where the establishment of territorial rights is excessively tangential to the establishment of jurisdictional authority. Claims to territorial rights in Antarctica are an example of claims that would violate this principle.*[70]

Furthermore, the presence of public officials (and buildings) for the benefit and control of the population may be one of the functions that gives states territorial rights, but it is a different thing when the presence of those officials and buildings is almost purely symbolic and fulfils almost no practical purpose. As for the regulation of an economic activity like whaling, one could say that stewardship requires control of the territory. But it is still left unjustified why one state rather than any other acquired rights to be the steward and grant licences in the first place.

The original motivation of function-based arguments is to justify why the common authority of a political community ought to have territorial rights. If guaranteeing peace and order, the establishment and enforcement of law, and the fulfilment of (at least) basic rights is what we want that authority to do for us, then granting it territorial rights seems not just desirable, but necessary. When arguing this way, recall that *who* belongs in the political community and *where* the physical borders of the political community lie are taken as given. But this is what the Antarctic case unsettles. The claimants could not argue that they needed control over the territory to fulfil their duties towards their subjects; and, if willing to be environmental guardians, they had to convince others (especially their potential competitors) why they were the best suited for the job. What happened in practice was that, absent any agreement among the claimants, they had to negotiate a treaty and 'freeze' the claims. In drafting the Antarctic Treaty, function-based arguments were again central, but this time they were used to justify the joint guardianship of a group of states over the whole continent, rather than the jurisdiction of individual states over individual slices of it.

[70] Nine, *Global Justice*, 42.

2.6 Concluding remarks

In this chapter, I presented the main arguments offered by the seven states that claimed Antarctic territory between 1908 and 1940, dividing them into two main kinds. On the one hand, connection-based arguments appealed to physical and symbolic links between the state or its nationals and the continent; on the other hand, function-based arguments underlined state activity and presence. I showed why neither could justify the huge extensions claimed and pointed out that the lack of agreement and recognition among themselves finally led the claimants to draft and sign the Antarctic Treaty.

The unmentioned elephant in the room constitutes the topic of the next chapter. Although connection and function were at the basis of states' reasoning, what led them to make claims in Antarctica was a deeply colonial mindset and logic, supported by geographical doctrines conveniently developed to justify takings of such magnitude. While most normative definitions of colonialism make the subjection of peoples its central moral wrong, I go further to show that colonialism with no people, as happened in Antarctica, was also morally problematic, albeit for other reasons.

3
Colonialism on Ice

Now, look at this, the latest chart which Europeans have drawn of this region. It is a blank, perfectly white

Henry Morton Stanley, Through the Dark Continent[1]

The quote above is from the US-Welsh explorer Henry Morton Stanley and refers to the heart of Central Africa, the area where he conducted most of his explorations and where he acted as colonial administrator during the second half of the nineteenth century. For Stanley, like for those drawing the 'New World' charts, places like Africa, Oceania, and North America were purposefully pictured as clean slates to be occupied by Western powers, 'enormous void[s] . . . about to be filled up',[2] ignoring and erasing the already existing populations and cultures that had long inhabited those places. What was so unsettling about Antarctica was that the encounter with the 'blank' was not a deceptive metaphor, but the plain reality.

Well into the nineteenth century, Antarctica was popularly imagined as being inhabited by 'Antipodeans', native people who lived in a temperate hollow in the South Pole.[3] According to historian Ben Maddison, the lack of indigenous human populations unnerved the newcomers, who were only met by penguins.[4] A vivid account of these first encounters with Antarctic 'natives' comes from Midshipman Poynter, member of a British naval expedition that arrived in January 1820 on what later came to be known as Penguin Island:

[1] Henry Morton Stanley, *Through the Dark Continent: Or, the Sources of the Nile, Around the Great Lakes of Equatorial Africa and Down the Livingstone River to the Atlantic Ocean*, vol. 2 (William Clowes and Sons, 1878), 195. I thank Joanne Yao for pointing me to this reference.

[2] Ibid.

[3] The theory of the 'hollow poles' was proposed in 1818 by the American writer John Symmes. In his only novel, Edgar Alan Poe describes the encounter between the protagonists and the inhabitants of the South Pole, an apparently friendly black tribe for whom the colour white is alien: Edgar Allan Poe, *The Narrative of Arthur Gordon Pym, of Nantucket Comprising the Details of a Mutiny* (Harper & Brothers, 1970).

[4] Maddison, *Class and Colonialism*, 52–53.

From Sovereignty to Guardianship. Alejandra Mancilla, Oxford University Press.
DOI: 10.1093/9780197902769.003.0003

> *While rounding the island we perceived the shore covered with penguins whose awkward motions made the most ridiculous appearance—while our noses were assailed with the most intolerable stench I ever smelt... It would appear incredible was I even to give a rough calculation of the multitude of penguins I met on landing... suffice it to add our progress was completely arrested by these gentlemen who with the most determined obstinacy disputed our right to proceed and it was not until great slaughter had been committed and an opening forced through them with lances, seal clubs & c we were enabled to further our research.*[5]

For Maddison, as well as for a growing number of researchers in the Antarctic humanities and social sciences, these quotes are a reminder that Antarctica was part and parcel of the last wave of global colonialism, and it is in that context that we must understand its early history.

In the last chapter, I reviewed how Antarctica entered this history, first and foremost as a treasure trove of natural resources open for anyone to exploit. I showed how the desire of some countries to secure those resources for themselves or to simply appropriate land led to a territorial race: between 1908 and 1940, seven states claimed different slices of the continent, leaving only the most inaccessible part unclaimed. I then presented the main arguments offered by states to justify their claims over Antarctic territory. I suggested that, while those arguments might serve to justify at best territorial rights over quite limited areas of Antarctica (some islands and coastal areas), not even in their most generous interpretation were they enough to support the actual extent of the claims. Towards the end of the chapter, I pointed out that the only way to make sense of the claims was by putting them in a colonial context. In what follows, I elaborate on this idea: there was colonialism in Antarctica, even though there were no humans directly colonialized in the process, and it was morally problematic.

I first examine how colonialism manifested itself in Antarctica through the application of three geographical doctrines that were eminently colonial devices: continuity, which allowed the claiming power to extend its territorial claim from a very limited area all the way into the *hinterland* (or, in this case, the *hintereis*); contiguity, which was the notion of continuity applied to lands that were contiguous rather than continuous, that is, divided by the sea; and

[5] Charles Wittit Poynter, *The Discovery of the South Shetland Islands: The Voyages of the Brig Williams 1819–1820 as Recorded in Contemporary Documents and the Journal of Midshipman C. W. Poynter*, ed. Richard John Campbell (Hakluyt Society, 2000), 111–14. Today, the island in question has an estimated population of 7,500 pairs of chinstrap penguins and 3,000 pairs of Adelie penguins.

the sector principle, a doctrine initially developed to sort out territorial claims in the Arctic in the form of pie slices projected from the outward borders of the adjacent states all the way to the pole. I then consider what is morally problematic about the use of these doctrines, even in the absence of colonialized peoples. To put it in Lockean terms (which are the terms implicitly used by the claimants themselves when justifying their takings), I suggest that, while colonization may arguably be done respecting the fair appropriation and non-waste provisos, the essence of colonialism lies in their systematic violation. To recall, the fair appropriation proviso holds that one may appropriate without prejudice to others so long as 'there is enough and as good' for them. The non-waste proviso refers to appropriating only what one can meaningfully use, leaving the remainder to others.[6] While this refers to colonialism as a practice, I go on to suggest that colonialism can also be seen as a logic, mindset, or attitude that manifests itself when the need to occupy space and to take, use, and secure resources turns into intentional, more or less institutionalized, greed.

This analysis fills a gap between two strands of literature. On the one hand, the idea that colonialism happened and still happens in Antarctica is not new among Antarctic scholars. In a pioneering 2006 article, Klaus Dodds depicts Antarctica as a site for post-colonial examination, a place to look for 'how systems of colonial domination . . . persist in the contemporary era'.[7] Thus, although there were no people to subject, the continent serves as a point of departure for evaluating practices associated with colonialism, 'such as mapping, surveying, and the subjugation of territory and non-human populations'.[8] Adrian Howkins has suggested that Antarctica was an example of settler colonialism, epitomizing 'the elitist, racist and exclusionary mentality of the settler colonial project . . . brimming with potential economic wealth but also without the usual entanglements of empire' (namely, the question of how to deal with the native populations).[9] Sanjay Chaturvedi has argued that the process of penetration in the polar regions should be seen as part of a much larger process of colonialism and imperialism starting first in Europe and unfolding throughout the world,[10] and Alan Hemmings has observed that Antarctic claims are 'grounded in unexamined late nineteenth- to early twentieth

[6] Locke, 'Second Treatise', §27 and §31, 288 and 290.

[7] Klaus Dodds, 'Post-Colonial Antarctica: An Emerging Engagement', *Polar Record* 42, no. 1 (2006): 59–70, 60.

[8] Dodds, 'Post-Colonial Antarctica', 61.

[9] Adrian Howkins, 'Appropriating Space: Antarctic Imperialism and the Mentality of Settler Colonialism', in *Making Settler Colonial Space*, ed. Tracey Banivanua Mar and Penelope Edmonds, 29–52 (Palgrave Macmillan, 2010), 31.

[10] Sanjay Chaturvedi, *The Polar Regions: A Political Geography* (Wiley, 1996), 39.

century conceptions of global organization and national dignity (even where the claimant position was not formalized until later) long overtaken in just about every other international context, and protected from critical analysis by the absence of resident people's imperative usual in decolonization contexts'.[11] A recent interdisciplinary volume on the topic suggests that the question for Antarctic scholars is no longer whether colonialism happened in Antarctica, but rather how it happened.[12] What most of these analyses lack, however, and what I attempt to provide here, is a more explicit account of the specific moral wrong of colonialism without people and, later in the book, a positive account that goes beyond the criticism and asks what may be done in response to it.

On the other hand, colonialism has become a topic of philosophical interest. Understood as a 'thick' concept (namely, one that has both descriptive and evaluative dimensions),[13] theorists have aimed at specifying the feature or features that make it morally problematic. For Lea Ypi, the wrong of colonialism lies in 'the creation and upholding of a political association that denies its members equal and reciprocal terms of cooperation',[14] where the members of that association belong to territorially distinct groups. Relatedly, Anna Stilz refers to the imposition of practices and institutions over the colonized in a way that affects their freedom as makers of their own rules.[15] Vittorio Bufacchi singles out the systematic use of violence aimed at domination,[16] while Margaret Moore focuses on the taking of land and the dispossession of native peoples, effectively impeding their ability to act as self-determining collectives.[17] Daniel Butt mentions three morally problematic features of colonialism: domination, exploitation, and cultural imposition.[18] What these conceptualizations of colonialism share is that the feature that makes it wrong has to do with the treatment of humans. Here, instead, I present a version of colonialism with no directly subjected humans, but which is nonetheless morally problematic. This helps to highlight a dimension of colonialism that tends to go unnoticed in humanly populated areas; namely, as a zero-sum game where the goal is to get as much land and as many resources as possible for oneself, with no consideration

[11] Alan D. Hemmings, 'Security Beyond Claims', 75.

[12] Peder Roberts and Alejandra Mancilla, eds., *Colonialism and Antarctica: Attitudes, Logics, and Practices* (Manchester University Press, 2024), https://doi.org/10.7765/9781526170644.

[13] Bernard Williams, *Ethics and the Limits of Philosophy* (Fontana Press/Collins, 1985), 129–30.

[14] Ypi, 'What's Wrong with Colonialism'.

[15] Anna Stilz, 'Decolonization and Self-Determination', *Social Philosophy and Policy* 32, no. 1 (2015): 1–24, https://doi.org/DOI: 10.1017/S0265052515000059.

[16] Bufacchi, 'Colonialism, Injustice, and Arbitrariness', 201.

[17] Margaret Moore, 'Justice and Colonialism', *Philosophy Compass* 11, no. 8 (2016): 447–61, https://doi.org/10.1111/phc3.12337.

[18] Daniel Butt, 'Colonialism and Postcolonialism', in *The International Encyclopedia of Ethics*, ed. Hugh LaFollette (Wiley-Blackwell, 2013).

for others, and even if the land and resources serve no immediate (or even mediate) purpose. Some may point out here that this is a description of capitalism rather than colonialism, and underline that Antarctica was a site where international capital flourished thanks to the exploitation of marine animals. While I do not deny this, I think that capitalism and colonialism in Antarctica were intertwined and mutually constituted.[19] But capitalism alone does not explain the claiming of the whole continent.[20]

Finally, an important clarification is that, for the purposes of this chapter, I take methodological anthropocentrism for granted. By this, I mean that I focus on the morally problematic features of Antarctic colonialism for other human beings and leave aside the question whether Antarctic colonialism also wronged non-humans and nature. Thus, the argument presented here should be persuasive even for those who do not accept that colonialism can wrong beings or entities other than humans.[21]

3.1 Colonialism or imperialism?

Before proceeding, it is important to clarify why I characterize the occupation of Antarctica as a case of colonialism rather than imperialism. After all, were the claimants not aiming at political and strategic aggrandizement, which seems to be the essence of the latter? And was it not the British Empire that started the territorial race on the continent? The definitions of colonialism and imperialism, and their relationship, are matters of ongoing contestation among scholars, and I do not aim to settle the discussion here.[22]

[19] The connections of capitalism and colonialism with regards to the sealing industry are explored in Cruz, Salerno, and Zarankin, 'The Case of Sealing', 200–2.

[20] Furthermore, the logic of hoarding land and resources is not exclusive to capitalism. Some authors have persuasively shown it to be at work under socialism too. See, for example, Roman Khandozhko, 'Antarctic Minerals for the Soviet Bloc? Imagining the South Pole Frontier of Extractive Socialism', in *Colonialism and Antarctica: Attitudes, Logics, and Practices*, ed. Peder Roberts and Alejandra Mancilla (Manchester University Press, 2024), https://doi.org/10.7765/9781526170644.00008 and Ryan T. Jones, *Red Leviathan: The Secret History of Soviet Whaling* (University of Chicago Press, 2022).

[21] For why nature should be decolonized and what this would imply in terms of global governance, see Alejandra Mancilla, 'Effective Deoccupation: Towards Responsible Guardianship of Nature', *Political Philosophy* 2, no. 2 (2025), https://doi.org/10.16995/pp.23671.

[22] See, for example, Margaret Kohn and Kavita Reddy, 'Colonialism', in *The Stanford Encyclopedia of Philosophy*, summer 2024 edition, ed. Edward Zalta and Uri Nodelman, https://plato.stanford.edu/archives/sum2024/entries/colonialism/; Krishan Kumar, 'Colony and Empire, Colonialism and Imperialism: A Meaningful Distinction?', *Comparative Studies in Society and History* 63, no. 2 (2021): 280–309; Barbara Arneil, 'Colonialism Versus Imperialism', *Political Theory* 52, no. 1 (2024): 146–76, https://doi.org/10.1177/00905917231193107; Barbara Bush, *Imperialism and Postcolonialism* (Routledge, 2014).

One way to look at the distinction between colonialism and imperialism is through etymology. Colonialism derives from the Latin *colonia*, a Roman settlement in newly conquered or hostile territory. Imperialism, instead, derives from the Latin verb *imperare*, to exercise dominion over others. While etymology is no knockout argument, colonialism is associated with the physical presence of the colonists in the colonial space (thereby 'settler' colonialism), but imperialism need not be. Colonialism is also associated, as opposed to imperialism, with land and resource appropriation, extraction, and exploitation—where 'resource' can be anything or anyone, non-human or human. In the words of Aimé Césaire, 'the decisive actors here are the adventurer and the pirate, the wholesale grocer and the ship owner, the gold digger and the merchant, appetite and force'.[23] Thus, while 'resource colonialism' is a category, 'resource imperialism' is not.[24] Under this distinction, then, colonialism denotes an instrumental way of relating to the external world—where the 'external world' may even include other humans. Imperialism, instead, is about the exercise of power over other humans, close and far. The two, of course, can and often are combined, like when an empire extends its presence through concrete colonial settlements, as happened with Rome in North Africa, Spain in Mexico, Central and South America, and the United Kingdom in India. But they can also be found independently, such as when the British settlers shook their subservience to the Crown and founded their own United States (arguably, colonialism without imperialism),[25] and when, later, the United States became an empire of its own, exercising power over others without the need to directly occupy their lands and extract their resources (arguably, imperialism without colonialism).

With this distinction at hand, I offer three reasons why one may think of Antarctica as a colonial site, without precluding the possibility that it might also have been an imperial one.

First, as was mentioned before, the original motivation for going to Antarctica was not ruling over the inhabitants but extracting its resources—which required some degree of occupation of the space. Antarctica, in this sense, was a clear example of 'resource colonialism', with seal hunters and

[23] Aimé Césaire, *Discourse on Colonialism*, trans. Joan Pinkham (Monthly Review Press, 2000), 33.

[24] On resource colonialism, see, for example, Andrew Curley, 'Resources Is Just Another Word for Colonialism', in *The Routledge Handbook of Critical Resource Geography*, ed. Matthew Himley, Elizabeth Havice, and Gabriela Valdivia (Taylor and Francis, 2021); Daria Gritsenko, 'Energy Development in the Arctic: Resource Colonialism Revisited', in *Handbook of the International Political Economy of Energy and Natural Resources*, ed. Andreas Goldthau, Michael F. Keating, and Caroline Kuzemko (Edward Elgar Publishing, 2018), 172–84.

[25] This is the argument in Arneil, 'Colonialism Versus Imperialism'.

whalers as the main agents in the process. Later, speculation about the hidden riches of the continent was the driving force behind much of the exploration and claim-making. The Australian explorer, Douglas Mawson, for example, claimed in 1915 that, 'from an economic aspect, the frozen South may not attract immediate attention. But who can say what a train of enterprise the future may bring?'[26] And Manuel Barros Recabarren, president of the Chilean War Academy in 1991, looked back on the establishment of Chilean presence in Antarctica as an opportunity that opened 'immense possibilities of development for our country, with incalculable riches hidden under the ice that will cement its progress and growth in the coming century'.[27]

Second, for many of those who were 'first' in Antarctica, Antarctica was not a special destination, but an intermediate stop in their colonial tour around the world. The seamless continuity of purpose, mindset, and practices is reflected in the chronicles they left. Just to give one example, Dumont D'Urville's *Two Voyages to the South Seas* is subtitled 'to Australia, New Zealand, Oceania 1826–1829 . . . and to the Straits of Magellan, Chile, Oceania, Southeast Asia, Australia, Antarctica, New Zealand and Torres Strait 1837–1840'.[28] In the second volume, a picture from the village square in Rajah Basa, Sumatra, appears right before a sketch of the rocks off Adélie Land. The explorer mentions that an advantage in claiming the latter is that it 'will never start a war against our country'[29]: here, the 'natives' could not offer much resistance.

Third, some of the explorers had an ambition to build a settler colonial Antarctica; they saw in the white, empty continent a promised land where a society unblemished by the corruption of temperate climates could flourish. This was Richard E. Byrd's dream while overwintering on the Ross Ice Shelf, in 1928–1929. As he wrote in his diary: 'In a word, we are trying to get away from the false standards by which men live under more civilized conditions'.[30] Domesticating the landscape to turn it into 'home' was also a common trope among some of the national expeditions, especially those of Argentina and Chile. In charge of the Chilean military station O'Higgins in 1948, Hugo Schmidt Prado reported that they had brought sheep, chicken, and dogs, and

[26] Douglas Mawson, *The Home of the Blizzard*, vol. 1 (William Heinemann, 1915), 6.

[27] Manuel Barros Recabarren, 'Presentación', in *¡Base O'Higgins Sin Novedad!*, second edition, by Hugo Schmidt Prado (Editorial La Noria, 1992), 14.

[28] Jules Sébastien César Dumont d'Urville, *An Account in Two Volumes of Two Voyages to the South Seas by Captain Jules S-C. Dumont d'Urville of the French Navy to Australia, New Zealand, Oceania 1826–1829 in the Corvette Astrolabe, and to the Straits of Magellan, Chile, Oceania, South East Asia, A*, vol. 2, trans. and ed. Helen Rosenman (Melbourne University Press, 1987).

[29] Ibid., 474.

[30] Richard Evelyn Byrd, *Little America: Aerial Exploration in the Antarctic and the Flight to the South Pole* (G. P. Putnam's Sons, 1930), 193.

that a visiting scientist was getting ready to spread the seeds of Fuegian plants in the barren ground, in the hope that they would grow there again as they had done millions of years ago.[31]

In sum, there was a motivation to appropriate Antarctic space and what was within it that went beyond the motivation of becoming more politically powerful through mere territorial expansion.

3.2 Colonial strategies

In his *Principles of Public International Law*, Ian Brownlie refers to three geographical doctrines that were developed in connection with the principle of effective occupation: continuity, contiguity, and the sector principle. Continuity referred to the idea that, when a state has permanently occupied a territory, it also acquires rights to the lands surrounding it. Contiguity was invoked when the claimed lands were divided by the sea, for example, islands facing the mainland. The sector principle, finally, was proposed as a way of delimiting territory in the polar regions (initially the Arctic) by drawing straight lines all the way to the pole from the extreme ends of the circumpolar territories. In Brownlie's words,

> *continuity and contiguity are a facet of the modern view of sovereignty according to which it does not depend on close settlement but on state activity. State activity as evidence of sovereignty need not press uniformly on every part of the territory.*[32]

As for the sector principle,

> *while the sector principle does not give title which would not arise otherwise, if the necessary state activity occurs, it represents a reasonable application of the principles of effective occupation as they are now understood.*[33]

Brownlie does not criticize these doctrines, but endorses them as 'logical and equitable principles which . . . come to the fore when the disputed territory is uninhabited, barren, or uncharted'.[34] What he omits from his analysis

[31] Hugo Schmidt Prado, *¡Base O'Higgins Sin Novedad!*, second edition (Editorial La Noria, 1992), 38.
[32] Ian Brownlie, *Principles of International Law*, seventh edition (Oxford University Press, 2008), 143.
[33] Ibid., 144.
[34] Ibid., 142–43.

is the historical context in which these doctrines emerged and flourished: it was by appeal to continuity that European powers could claim the *hinterland* of the areas they occupied along the African coast during the second half of the nineteenth century. And it was by assuming contiguity and, in some cases, the sector principle, that the Arctic was divided among the countries adjacent to it during the first decades of the twentieth century. These three doctrines, in sum, can only be understood within the framework of late global colonialism and its accompanying determination to take areas much vaster than what states were able to meaningfully control and occupy.

The application of these doctrines shines in all its glory in Antarctica, with the continent divided in orderly, continuous slices from the pole to the coasts and nearby islands and archipelagos. In fact, all claimants but Norway claimed sectors that stretched to the South Pole.[35] Even Argentina and Chile, the two countries that stood against the others as purportedly anti-imperial and anti-colonial, explicitly mentioned the use of these geographical doctrines in the shaping of their claims.[36] As seen in Chapter 2, moreover, the South American claimants appealed to the Treaty of Tordesillas as giving them historical rights over Antarctica, although in 1494 the continent had been sighted only in the imagination. In order to defend their position as anti-colonial and anti-imperial, ironically, they invoked a colonial treaty par excellence.

To be clear, the claimants used these doctrines not as roots of title (i.e. as giving them original rights over the continent) but as devices to extend their

[35] Although Norway left the southern limit of its Antarctic territory undetermined in its original claim, in 2015 a White Paper from the Ministry of Foreign Affairs mentioned for the first time that its claim went all the way to the South Pole: Ole Magnus Rapp, 'Norge Utvider Dronning Maud Land helt frem til Sydpolen', *Aftenposten*, 19 September 2015, http://www.aftenposten.no/norge/Norge-utvider-Dronning-Maud-Land-helt-frem-til-Sydpolen-28019b.html.

[36] The Argentinians claim that 'when a state has permanently occupied a territory, it acquires rights to occupy other lands surrounding it. This has been called the "doctrine of continuity". If a stretch of the coast were occupied, for example, the sovereignty of the state can be extended to the islands and lands close by, in which case the "doctrine of contiguity" would be involved'. (Comisión Nacional del Antártico, *Soberanía Argentina en la Antártida*, second edition (Ministerio de Relaciones Exteriores y Culto, 1948), 65.) Oscar Pinochet de la Barra gives four reasons why it makes sense to apply the sector principle in the Chilean case: '1. The presumption or belief that a country adjacent to an Antarctic territory has therein exercised, in permanent or almost permanent form, acts of dominion, with more frequency and political intention than another that is non-adjacent. 2. The necessity for an adjacent country of possessing the Antarctic sector facing it, because of its national defence, the establishment of meteorological observatories for the development of its aviation, of its agriculture, etc. 3. The possibility of exploiting its natural resources more properly. 4. The impossibility of delimiting rights over an immense, unoccupied, inaccessible continent, except by meridians that form a triangle with its vertex at the Pole—the occupation of certain accessible points along the coast being sufficient': Pinochet de la Barra, *Chilean Sovereignty*, 46–47. For an argument against the idea that the claims from Chile and Argentina were colonial or imperial, see Ignacio Javier Cardone, 'Argentina and Chile's Antarctic Colonialism? A Postcolonial Critique to Eurocentric Analysis', in *Colonialism and Antarctica: Attitudes, Logics and Practices*, ed. Peder Roberts and Alejandra Mancilla, 99–121 (Manchester University Press, 2024), https://www.manchesterhive.com/display/9781526170644/9781526170644.00011.xml.

territories from the areas where they had some sort of presence. Yet these doctrines are morally problematic. Today, when colonial models of appropriation have come under scrutiny, this statement seems obvious, but even at the time when they were used there were critical voices. One of them was the Norwegian jurist Gustav Smedal. Contrary to those who advocated for a relaxation of the requirements of effective occupation in the polar areas, he argued in 1931 that:

> *The fact that it may be difficult, or in some cases perhaps impossible, at present to take effective possession of a polar land, does not warrant a disregard of the rule in international law relating to occupation. The land must, in that event, continue to be unoccupied. No stipulations exist to the effect that every land shall be submitted to sovereignty, and neither is there any need for such stipulations.*[37]

Smedal did not go as far as suggesting that, for uninhabited and mostly uninhabitable places like Antarctica, one should think about the question of occupation and governance in an altogether different manner—which is what I go on to suggest later in this book. For him, the issue was more about leaving as *terra nullius* what may not be effectively taken and occupied. Smedal was not in principle against the idea of states making sovereign claims in the Arctic or Antarctic, but against the procedure that was being used to make them.[38] He agreed that land and non-human natural 'resources', living or not, were there to be appropriated and exploited by humans through nation-states. What he objected to were the terms used to allow takings of a scale and at a pace never seen before in the history of humankind. In other words, Smedal's critique assumed methodological nationalism (states were the paradigmatic territorial agents) and methodological anthropocentrism (beyond humans, everything and everyone is a potential resource). However, even within that framework he saw that the Antarctic claims were excessive.

[37] Smedal, *Acquisition of Sovereignty*, 61.

[38] The following quote makes this clear. In the case of the Arctic, he says, 'the parties on whom the greatest wrong would be inflicted by the sector principle are the states that are not bound by the Arctic Sea. Any state whatsoever may, from scientific or economic reasons, be interested in having the sovereignty over an Arctic land, and it is quite illegitimate to exclude such a state from obtaining this on the pretence that its territory is not lying sufficiently far to the north. Lakhtine objects to this view on the ground that the interests of these states in the Arctic can only be of an "imperialist character," and that the interests for this reason "cannot be recognized as being reasonable." However, it cannot in any way be admitted that a sector state, in looking after its economic and political interests in the Arctic, is performing an act of a more elevated or ideal character than any other state does in looking after its interests': ibid., 62. In the case of Antarctica, the application of the sector principle is even more arbitrary than in the Arctic, considering that some of those claiming territory are not even adjacent or contiguous states.

Other opponents of the application of the sector principle in the Arctic also argued in a way that easily applies to Antarctica. Again, they were not questioning that the earth was there for humans (through states) to take, but they doubted the morality of legal doctrines that allowed states to appropriate huge extensions of land and the resources in them purely by letter. As summarized by Ivan L. Head:

> *National claims under the [sector] theory were in exact reverse order to the normal process of acquisition. The sector theory places territory in the legal possession of a state even before it is discovered. How, it was argued, can a state claim sovereignty over areas about which it knows absolutely nothing?*[39]

Or as T. E. M. Kitterick put it, more dramatically:

> *If the title conferred by mere contiguity is not inferior to that conferred by actual occupation, where is the boundary to be drawn? A state might almost claim sovereignty over a whole continent after establishing a single settlement.*[40]

One lesson to draw from this is that Antarctic claimants were not alone in abusing colonial geographical doctrines. If we have lost sight of this aspect of colonialism—the hoarding of land and resources by states, sometimes merely by drawing lines in a map—it is because the injustices committed against other humans have been the focus of attention. What the Antarctic case allows us to see is that, even if one grants methodological nationalism and methodological anthropocentrism, colonialism without directly colonialized people remains problematic.

3.3 Colonialism versus colonization

The use of the geographical doctrines of continuity, contiguity, and the sector principle, and the appeal to a colonial treaty like Tordesillas to claim Antarctic territory, reveal what is wrong with colonialism without people: even if one conceded that all land in the world should eventually be governed by states and be exploited by humans and for humans (i.e. even if one did not question

[39] Ivan L. Head, 'Canadian Claims to Territorial Sovereignty in the Arctic Regions', *McGill Law Journal* 9, no. 3 (1963): 200–26, 205.
[40] T. E. M. McKitterick, 'The Validity of Territorial and Other Claims in Polar Regions', *Journal of Comparative Legislation and International Law* 21 (1939): 89–97, 92.

methodological nationalism or methodological anthropocentrism), one could still reject Antarctic claims on the grounds that they far exceeded what was reasonable to claim in any meaningful way. (As I show in Chapter 5, if one questions these biases there are still further reasons to reject colonialism in Antarctica, the most obvious being the exploitation of non-human animals and their treatment purely as resources.)

Some may object that what I am criticizing here is what humans have done throughout history, namely, to colonize new spaces and bring them under their control. However, this objection fails to distinguish colonization from colonialism. Unfortunately, the English language does not help to make this distinction: in the *Oxford English Dictionary*, the first three definitions of the verb 'to colonize' all refer to settling and dominating in a colonial manner, while only the fourth brings forth what I wish to underline here, namely, the more neutral sense of the word.[41] In this sense, to colonize means to occupy the space around us and take, use, and secure the resources around us in order to live. Bacteria, fungi, plants, and animals including us have always colonized and will keep colonizing; this is what it takes to be a functional living organism. I propose to understand colonialism, instead, as an intentional and somewhat institutionalized human activity aimed at taking, using, and securing land and resources (living and non-living, non-human and human) in a way that greatly exceeds the ends of colonization.[42] While I use the verb *colonize* and the adjective *colonizers* as neutral terms, I understand the verb *colonialize* and the adjective *colonialists* as 'thick' ones: they describe a way of behaving in the world and provide an implicit (negative) valuation of it.[43]

This distinction is useful, I submit, because it allows us to understand that Antarctica has to date been minimally *colonized* by humans, but almost entirely *colonialized*; that colonialism can be to some extent performed from the armchair, while colonization is always hands-on;[44] and that colonialism

[41] 'To spread into or become established in (an area, habitat, host organism, etc.); to form a colony in (a place) or on (a substrate or host)': Oxford English Dictionary, 'colonize, (*v.*)', March 2025, https://doi.org/10.1093/OED/4417620177.

[42] I say 'somewhat institutionalized' because, as some scholars of colonialism have noted, colonial processes are normally much less planned and more chaotic than what is normally thought. For one example, see William Dalrymple, *The Anarchy: The Relentless Rise of the East India Company* (Bloomsbury Publishing, 2019).

[43] Are species like honey fungus not colonialist too, spreading along whole forests when they go unchecked? So long as we are not smart enough to communicate with other species and understand their intentions (if they have any), I leave these questions open. It might turn out that colonialism is not exclusive to humans; only that we have been the most successful at it.

[44] Note that, when performed from the armchair, colonialism can look very similar to imperialism. However, colonialism requires some dimension of 'fieldwork'—at least an initial sign of human presence wherefrom borders on the map can later be drawn.

in Antarctica was exercised as shamelessly as anywhere, in the absence of any previous human colonizers to be colonialized—opposite to what was the case in the rest of the world, where colonialists had to confront those human groups already colonizing the claimed spaces.

Bringing back the Lockean analysis of land and resource appropriation presented in Chapter 2, one could say that while colonization may be done in a non-wasteful manner and respecting the fair appropriation proviso, the wrong of colonialism lies in the systematic violation of both. As shown before, in the justification for their Antarctic takings, the claimants appealed to various ways in which they mixed their labour with the land and added value to it. While this may have been a reasonable justification to claim very moderate pieces of coastal land and maybe some Antarctic and sub-Antarctic islands, it is not so when applied to the claims as they stand. The claims could not have been made, thus, without assuming that no one was being deprived as a result (namely, that enough and as good was being left for others), and that the claimants were not 'wasting' the land, in the sense of removing it from the commons for no good reason. But both these assumptions are contestable. Antarctica was the last big expanse of land to be claimed by states, and it constituted their last opportunity for territorial aggrandizement at the expense of others and in direct competition with others (recall, for example, the historical rivalry between France and the United Kingdom).[45] Furthermore, it should not be controversial to claim that what they did was wasteful; a mere hoarding of land without any value-adding in the process. Through their actions, the claimants precluded not just the possibility of others making independent claims over Antarctica as *terra nullius*; they also precluded the possibility of keeping it as a *terra communis* beyond the jurisdiction of individual states. As I show in Chapter 4, absent any agreement among the claimants, and between the claimants and the rest of the world, Antarctica became *per force* a unique kind of territory. But this was more an inevitable result than something actively sought by the parties.

While the most egregious forms of colonialism have been connected to the direct subjugation of other humans and their treatment as resources rather than equals, the Antarctic case shows that colonialism can happen without directly subjugated people and still be wrong. Margaret Moore underlines that what is wrong with colonialism is the unjust taking of other people's land and resources.[46] In a twist to Moore's account, one could say that the wrong of Antarctic colonialism lies not in the taking of land and resources from others

[45] That is, at least on earth. Outer space is the obvious next ground for colonialism.
[46] Moore, 'Justice and Colonialism', 455.

who had already colonized them, but in the taking of land and resources with no good purpose from the commons.

Some may wonder yet why I choose Locke's theory as a reference point. After all, his account of initial appropriation and, indirectly, territorial rights, relies on a questionable belief in natural rights and has been criticized on many fronts, including that of being an apology for settler colonialism with all its associated human rights violations, from political domination to slavery.[47] My answer to this is straightforward: if not even a theory that has been instrumental in the justification of colonialism can justify the occupation of Antarctica, then there are no moral arguments left for colonialists aiming to justify their hoarding of land and resources. Beyond this point, territorial claims of this kind can only be justified by appeal to *Realpolitik* or brute force—which is not something that the Antarctic claimants ever openly admitted, and which is something that other colonial states using similar procedures should also begin to acknowledge.

So far, I have focused on the wrongs of colonialism as a practice, a way of acting in the world, and its negative effects on others. But one could also make a normative judgement of colonialism as a logic, a mindset, an attitude. Distinguishing between colonization and colonialism can again help to make this clear: while colonization is driven by need (needs that all living beings share), colonialism is driven by greed. Under the conditions of moderate scarcity that David Hume deemed necessary for justice and injustice to arise, this distinction is easy to see.[48] However, one must not lose sight of it when looking at Antarctica and at other places that came to be under other states' jurisdictions based on this same logic.

3.4 'The Antarctic Problem'

In 1908, the United Kingdom made the first territorial claim over Antarctica, through a hasty Letters Patent that had to be amended in 1917. France, New Zealand, Australia, Norway, Chile, and Argentina followed, with the claims of the last two mostly overlapping with the British one in the area of the Antarctic Peninsula. The United States and the Soviet Union adopted a different strategy: they refrained from making claims, but did not recognize the others'

[47] See, for example, Barbara Arneil, *John Locke and America: The Defence of English Colonialism* (Clarendon Press, 1996).
[48] David Hume, *A Treatise of Human Nature*, vol. 2 (Everyman's Library, 1974), III.II.2, 199.

claims and left open the possibility that they might make their own claim in the future. During the 1940s and especially after the end of World War II, the 'Antarctic Problem' became pressing: while some of the claimants accepted the legitimacy of each other's claims, the two 'potential' claimants accepted none, and the three overlapping claimants started a race towards the south to display their sovereignty. This race appealed to strategies that resembled those of elementary school children more than those of grown-up states. During the secret Operation Tabarin, from 1943 to 1946, the United Kingdom established permanent stations, issued stamps from the Falkland Islands Dependencies (of which the Antarctic claim was part), and conducted scientific research 'to raise Britain's profile in the South Atlantic and Antarctic'.[49] There was soon a South American counteroffensive. In January 1947, at Melchior Base on Melchior Island (Palmer Archipelago, Bellingshausen Sea), the Argentinians arrived and removed a plaque that the British had affixed to the lighthouse, repainted it with the Argentinian national flag colours, and added the inscription: 'Property of the Marine Ministry of the Republic'.[50] In February 1948, on Deception Island, Brits and Argentinians met and tried to resolve the tension with a football match, with the Brits winning by 1–0.[51] In March 1948, it was the Chileans' turn to encounter a British ship near Greenwich Island, where they had a base. The Chilean promised a welcome using 'a machine gun and six rifles', but ended up inviting the British for tea, while also rejecting the written protests the Brits had brought.[52] In 1956, the United Kingdom took the disagreement with Argentina and Chile to the International Court of Justice, but the two countries rejected the court's jurisdiction in the matter. The issue remained unresolved.

It is tempting here to imagine other possible worlds for Antarctica. If the claimants had agreed among themselves, would the continent have been divided like the rest of the world, in tidy slices among the countries that had an interest in it at the time? Would they be performing sovereignty there as elsewhere, settling populations, extracting resources, and controlling their borders? What if the tensions had gotten worse, to the point of starting a war for Antarctica? What would the Peninsula look like today, and what would have happened with the rest of the continent? What if the 'potential' claimants had taken a more aggressive stance and declared the place to be theirs instead?

[49] Dodds, *Pink Ice*, 14.
[50] Pablo Fontana, *La Pugna Antártica: El Conflicto por el Sexto Continente 1939–1959* (Guazuvirá Ediciones, 2014), 181–82.
[51] Ibid., 191–92.
[52] Ibid., 202.

Would there have been an East and a West Antarctica divided by an ideological wall?

What happened instead was that, absent a mutual recognition of the legitimacy of their claims, the claimants met to discuss a peaceable way out of the problem that would still be satisfactory for them. At the initiative of the United States, they met several times during 1959 in Washington, in secret, and drafted a way forward. The options of condominium and internationalization were discussed but quickly dismissed. Under *condominia*, two or more states equally exercise sovereignty by means of an autonomous local administration.[53] Under internationalization, Antarctica would have been under the responsibility of the United Nations. Both alternatives seemed unacceptable to the claimants. Instead, in an exercise of creative political imagination, a *sui generis* solution proposed by the Chilean legal scholar Julio Escudero was finally accepted by all. This was captured in Article IV of the Treaty:

> 1. *Nothing contained in the present Treaty shall be interpreted as:*
>   a) *a renunciation by any Contracting Party of previously asserted rights of or claims to territorial sovereignty in Antarctica;*
>   b) *a renunciation or diminution by any Contracting Party of any basis of claim to territorial sovereignty in Antarctica which it may have whether as a result of its activities or those of its nationals in Antarctica, or otherwise;*
>   c) *prejudicing the position of any Contracting Party as regards its recognition or non-recognition of any other State's rights of or claim or basis of claim to territorial sovereignty in Antarctica.*
> 2. *No acts or activities taking place while the present Treaty is in force shall constitute a basis for asserting, supporting or denying a claim to territorial sovereignty in Antarctica or create any rights of sovereignty in Antarctica. No new claim, or enlargement of an existing claim, to territorial sovereignty in Antarctica shall be asserted while the present Treaty is in force.*[54]

By virtue of Article IV, the colonial claims were thus not given up entirely, but suspended, or, more appropriately in the Antarctic context, frozen. The 'potential' claims were left frozen too, and the possibility of changing existing and potential claims or of making new ones was foreclosed. Furthermore, intended as an ecumenical document, the treaty was open for actual claimants, potential

[53] Brownlie, 'International Law', 60; Malcolm N. Shaw, *International Law*, eighth edition (Cambridge University Press, 2017), 179–80.

[54] Conference on Antarctica, 'The Antarctic Treaty', 1 December 1959, http://www.ats.aq/documents/ats/treaty_original.pdf.

claimants, and non-claimants, regardless of whether they recognized the existing claims or not. One way of analysing this piece of diplomatic engineering is to say that, by freezing the claims, the Antarctic Treaty 'rewarded' and preserved colonialism. In Klaus Dodds' words:

> *Britain and the other claimant states were condoned as colonial powers for the sake of enabling agreement with others such as the United States and the Soviet Union who wanted to reserve the right to make their own claims in the future.*[55]

I say more about this claim in Chapter 4 but, in short, I agree with Dodds that Antarctica remains a place to be decolonialized and that the Treaty lost this opportunity. As time passes, it becomes more and more evident that those claims are anachronic and ultimately untenable.[56] At the same time, the Antarctic Treaty can also be seen as a one-of-a-kind agreement where, driven by the circumstances, a group of states committed to undertake Joint Guardianship of a whole continent without deriving any immediate benefits from it. They might have done this grudgingly; they might have done this with the hope of future gains. But the point remains that they created something remarkable: a new kind of territorial governance that followed neither the traditional sovereignty model as exercised by states, nor the international doctrine of Common Heritage of Mankind devised for places like the High Seas and the deep seabed. As I argue in Chapter 7, this model could be an inspiration for other places on earth that share important similarities and challenges with Antarctica.

3.5 Concluding remarks

In this chapter, I proposed the idea that territorial claims in Antarctica are only understandable in the context of the last wave of colonialism worldwide, with a few states appropriating for themselves vast swaths of land through geographical doctrines developed specifically for that purpose: continuity, contiguity, and the sector principle. I showed why colonialism devoid of people was still morally problematic: these takings were unreasonable and unjustifiable even by the standards used by the claimants, something that was reflected by the fact that they could not agree among themselves on how to divide the

[55] Dodds, 'Post-Colonial Antarctica', 63.

[56] I develop the idea that Antarctica remains to be decolonized in Alejandra Mancilla, 'Decolonising Antarctica', in *Philosophies of Polar Law*, ed. Alan Hemmings and Dawid Bunikowski (Routledge, 2021). See also Hemmings, 'Security Beyond Claims'.

pie. While this could have led to tensions and strife, however, by freezing their claims through the Antarctic Treaty, they achieved a unique result that meant preserving Antarctica rather than submitting it to the same regime of use and exploitation as the rest of the world.

In the next chapter, I examine how the Antarctic political community was constituted and what its self-appointed representatives took to be their main mission. Appealing to what is known as the *All Affected Interests Principle* in democratic theory, I suggest that criticism to the Antarctic Treaty arose when its parties envisaged the possibility of exploiting Antarctic mineral resources. By foreclosing that possibility with the signature of the Environmental Protocol, the parties kept Antarctica as a unique territory devoted to peace, science and—now with explicit emphasis—the protection of the environment.

4
Constituting the Antarctic Political Community

This chapter addresses the question of how Antarctica's political community and representatives were constituted through the Antarctic Treaty. While the problem of 'delimiting the demos' in political philosophy has focused entirely on how to bound already-existing political communities in already-inhabited territories, here we have a case where a human demos is constituted *for* a territory with no native humans in it. As I show later in the book, the answer given allowed, in my view, for a unique type of governance that may serve as a blueprint in other places.

An important clarification is that what follows is not an account of how the participants in Antarctic governance and their critics explicitly framed their positions, but rather my own interpretation of what they meant, using the available tools of political philosophy. This requires some simplification that, I hope, will not make those given to historical detail too impatient. There are also three starting assumptions that should be spelled out. First, I assume that territorial governance within and beyond universally recognized areas of state jurisdiction ought ideally to be decided through a democratic process. Second, by a democratic process I mean one that genuinely allows for the representation (if not direct participation) of those who would be relevantly affected by it. Even if it might be impossible to *constitute* a demos democratically, once *formed* the demos should constantly adjust to include those relevantly affected. Third, I consider the question of political representation to go hand in hand with that of distributive justice. It may sound strange to proceed this way in a place where the immediate, obvious issue to be resolved seems to be how to divide, or not divide, the land and resources. My point, however, is that to assess the fairness of any arrangement regarding the distribution of land and resources (including the question of whether to divide them at all), those whose relevant interests are potentially at stake in the distribution should be represented in the decision-making process.

From Sovereignty to Guardianship. Alejandra Mancilla, Oxford University Press. © Alejandra Mancilla 2026.
DOI: 10.1093/9780197902769.003.0004

I start by referring to what is known as the *boundary problem* in political philosophy, and by presenting what I see as the most plausible principle to address it, namely, the *All Affected Interests Principle*. I then show how the demos and its representatives were constituted for Antarctica by virtue of the Antarctic Treaty, and how this arrangement worked until the possibility of mineral exploitation began to be discussed. I look at critiques of the Antarctic Treaty through what was known as the 'Question of Antarctica' at the United Nations. I suggest that, by freezing the interests at stake (i.e. by freezing the possibility of anyone immediately profiting from the continent's riches), the Environmental Protocol kick-started a new era of Antarctic territorial governance, with environmental protection at its centre. I conclude that the Treaty and its critics did not question methodological nationalism and methodological anthropocentrism. However, the Protocol contains the tools to move beyond both, even though those tools have not so far been used for this purpose.

4.1 Constituting the demos

The question of bounding or constituting the demos has been much discussed in political theory. In Robert A. Dahl's words:

> *how to decide who legitimately make up 'the people'—or rather, 'a people'—and hence are entitled to govern themselves in their own association is a problem almost totally neglected by all the great political philosophers who write about democracy.*[1]

Or, as Frederik Whelan puts it:

> *Democratic theories usually focus on the internal decision-making arrangements, taking for granted the prior existence of a well-defined group with respect to which the question of democratic governance arises.*[2]

In answering the challenge above, some—like Whelan himself—have suggested that there is no principled way of setting the boundaries of the political community by democratic means, since these must be in place before

[1] Robert A. Dahl, *After the Revolution? Authority in a Good Society* (Yale University Press, 1990), 46.
[2] Frederick G. Whelan, 'Prologue: Democratic Theory and the Boundary Problem', *Nomos XXV: Liberal Democracy* 25 (1983): 13–47, 15.

democracy starts operating.[3] Others have appealed instead to a commonsensical principle, the application of which, however, has radical implications for our current understanding of democracy and democratic practice. According to the *All Affected Interests Principle* mentioned by Dahl and defended by, among others, Robert E. Goodin, all those whose interests would be relevantly affected should be part of the decision-making group, and democracy should be the preferred system to bring about those decisions.[4] In essence, the point of the *All Affected Interests Principle* is to recognize that, especially when living closely together, we affect one another. Insofar as this is the case, it is important that all those who have morally relevant interests that would be positively or negatively affected by some decision should have a chance to have a say in that decision. For Goodin, this explains why the existing demoi have generally been constituted based on a shared nationality, 'blood and soil', or a common history. These factors are not relevant in themselves, but by proxy: underlying them is the fact of intertwined (and therefore potentially affected) morally relevant interests.[5]

What makes the *All Affected Interests Principle* a radical departure from our common understanding and practice of democracy is the fact that, while intertwined interests are most obvious under conditions of physical proximity, physical proximity is not necessary for having one's interests intertwined. An obvious example is how a tariff established in one country can determine the economic prospects of producers in other countries; another, more causally muddled but no less certain, is how the CO_2 emissions of country M can end up putting country N's land underwater—no matter how environmentally mindful N has been. Yet another example regards migration policies and how they determine the prospects of millions of potential immigrants.[6] In all these cases, for the *All Affected Interests Principle* to be realized, everyone whose interests would be affected, no matter where, ought to be considered: a genuine democracy should not be constrained by geographical considerations, then,

[3] Ibid., 41–42.

[4] As Goodin and Arrhenius put it, '"would be" should be read as "would be, if the law were to be enacted"': Robert E. Goodin and Gustaf Arrhenius, 'Enfranchising All Subjected: A Reconstruction and Problematization', *Politics, Philosophy & Economics* 23, no. 2 (2024): 125–53, 126, https://doi.org/10.1177/1470594X241232023. See also Robert E. Goodin, 'Enfranchising All Affected Interests, and Its Alternatives', *Philosophy & Public Affairs* 35, no. 1 (2007): 40–68; Carol C. Gould, 'Structuring Global Democracy: Political Communities, Universal Human Rights, and Transnational Representation', *Metaphilosophy* 40, no. 1 (2009): 24–41, https://doi.org/https://doi.org/10.1111/j.1467-9973.2009.01568.x; and David Held, *Cosmopolitanism: Ideals and Realities* (Polity, 2010), 72.

[5] Goodin, 'Enfranchising All Affected Interests', 48.

[6] On the bounding of the demos and migration policies, see Arash Abizadeh, 'Democratic Theory and Border Coercion: No Right to Unilaterally Control Your Own Borders', *Political Theory* 36, no. 1 (2008): 37–65, http://journals.sagepub.com/doi/abs/10.1177/0090591707310090.

but by the commitment to include in its decision-making process all affected parties. This and no less ought to be the democratic ideal.

This is not the place to review the many responses to and critiques of the *All Affected Interests Principle.*[7] For our purposes, what matters is that the bulk of the discussion has focused on the constitution of the demos *in* humanly inhabited territories, paradigmatically states. Here, however, we are confronted with a different question; namely, who should be the demos *for* a humanly uninhabited and mostly uninhabitable place like Antarctica, when that place is brought under human jurisdiction? To answer this question, I suggest that the *All Affected Interests Principle* seems especially appropriate. Because the norms chosen to rule Antarctica relevantly affect others beyond Antarctica, it seems uncontroversial that these others should somehow be considered in the rule-making process. Moreover, here geographical space cannot be the pre-political or extra-democratic factor delimiting the demos, since neither those ruling nor most of those ruled inhabit it.

In what follows, I distinguish three different kinds of groups that are usually clumped together in the discussions of how to delimit the demos. First, there are all those who should be part of the political community, that is, all those who should be represented. This is the *demos* proper. Second, there are those who should elect the representatives of the political community. I call these the *enfranchised*. Third, there are those who should have the decision-making power to create norms that advance (or do not oppose) the best interests of the represented. These are the *representatives* (usually a sub-group of the enfranchised). In the existing literature, these three groups are for the most part assumed to be composed of human individuals, constituting a nation-state. This is not surprising, given that methodological anthropocentrism and methodological nationalism have largely been taken as default. However, when one moves to the level of international or global governance (as is the case with Antarctica), this changes. At this level, the demos is composed of individuals, albeit only very indirectly, while the enfranchised and the representatives are states, impersonated by human delegates.[8] As will be seen, there has never

[7] The main contender is the *All Subjected Interests Principle*, which claims that those *subjected* to the decisions made by a democratic polity (rather than those affected) should constitute the demos. See, for example, Ben Saunders, 'Defining the Demos', *Politics, Philosophy & Economics* 11, no. 3 (2012): 280–301, https://doi.org/10.1177/1470594X11416782; Sarah Song, 'The Boundary Problem in Democratic Theory: Why the Demos Should be Bounded by the State', *International Theory* 4, no. 1 (2012): 39–68, https://doi.org/10.1017/S1752971911000248; Carmen E. Pavel, 'Boundaries, Subjection to Laws, and Affected Interests', in *The Oxford Handbook of Freedom*, ed. David Schmidtz and Carmen E. Pavel (Oxford University Press, 2018), https://doi.org/10.1093/oxfordhb/9780199989423.013.11.

[8] It is no wonder that the democratic deficit of global governance is a focus of criticism. See, for example, Jonathan Kuyper, 'Global Governance and the Democratic Deficit', *Alternative Law Journal* 35 (2010): 169–70.

been direct participation of civil society in Antarctic deliberation; at most, there have been non-governmental groups pushing for greater inclusion of some issues, all done in a rather symbolic manner. States and states exclusively have been the main players.

4.2 Constituting the Antarctic demos and its representatives

> *A plunge into the writhing storm-whirl stamps upon the senses an indelible and awful impression seldom equalled in the whole gamut of natural experience. The world a void, grisly, fierce and appalling. We stumble and struggle through the Stygian gloom, the merciless blast—an incubus of vengeance—stabs, buffets, and freezes; the stinging drift blinds and chokes.*
>
> *Douglas Mawson, The home of the blizzard*[9]

Why would anyone care to impose their jurisdiction over a place as uninviting to human life as the quote above suggests? As shown in Chapter 2, during the first half of the twentieth century, when the territorial claims over Antarctica were made, the answer was straightforward: if you didn't, someone else would—where the underlying assumption was that having more land and natural resources was always better than having fewer, never mind the climate. States at the time were racing for the last bits of *terra nullius*, land not yet under anyone's territorial jurisdiction. As suggested in Chapter 3, Antarctica thus gave steam to a last phase of colonialism, with seven states using the geographical doctrines of continuity, contiguity, and the sector principle to claim as much land and as many resources as possible.

The answer to the above question changed when it became clear to the claimants and potential claimants that their claims were contestable, and when the extension of the geopolitical tensions of the Cold War to Antarctica became a real possibility. Figuring out a commonly accepted system of territorial governance among those interested in the place was necessary to preserve global peace, but they wanted to do it without entirely losing their presence and influence over Antarctica. If any one of those interested had unilaterally enforced their claim, conflict—maybe on a big scale—would have followed. Sitting down with the delegates from the other parties and negotiating was therefore deemed the way to go.

9 Mawson, *Home of the Blizzard*, 120.

What resulted was the Antarctic Treaty, signed in Washington on December 1, 1959, and ratified in 1961. The twelve initial signatories were the claimants (Argentina, Australia, Chile, France, New Zealand, Norway, and the United Kingdom), the two 'potential' claimants (the United States and the Soviet Union), and Belgium, Japan, and South Africa. What united them was that they all had been active participants in the International Geophysical Year of 1957–58 in Antarctica, building research stations there.[10] Taking advantage of the context, science (rather than the impossibility of legitimizing their territorial claims) was erected as the cornerstone of the Treaty, which was done—as stated in the preamble—'in the interest of all mankind'.[11] This was done by establishing the importance of freedom of scientific investigation and the promotion of international cooperation in scientific pursuits. Moreover, the performance of 'substantial research there, such as the establishment of a scientific station or the dispatch of a scientific expedition', became the criterion to determine who had decision-making power and who did not.[12] This resulted in a hierarchical division between consultative parties (members who conducted science on the continent and therefore had decision-making power) and non-consultative parties (members who did not fulfil the science criterion; they could participate in the Antarctic Treaty meetings but had no decision-making power). Thus, even though the Treaty was open for all states to join, science became the calling card for political influence.

In the language of democratic theory, the Antarctic demos was 'mankind' as a whole. Those representing it were the original group of self-enfranchised states plus newcomers who, to join the club, had to pass the science test to the satisfaction of the existing members. The criterion to become a representative was scientific activity, interpreted not just as having the right intention and curiosity but also the actual means for the performance of substantial research on the continent. Once the science test was passed, the full range of privileges followed: the power to formulate and adopt measures, decisions, and resolutions by consensus; decide who else should be allowed as a consultative party; send observers to Antarctic stations; convene the parties for a revision of the Treaty (and, later, of the Environmental Protocol); and retain the status of a consultative party *ad eternum*, regardless of whether substantial research

[10] The International Geophysical Year took place between July 1957 and December 1958 and included sixty-seven countries. One of the key spots was Antarctica.

[11] Conference on Antarctica, 'The Antarctic Treaty', Preamble, 1 December 1959, http://www.ats.aq/documents/ats/treaty_original.pdf.

[12] Conference on Antarctica, 'The Antarctic Treaty', Article VIII.

activity was kept up or not.[13] Unlike with most paid subscriptions, thus, here the 'payment' was required only once.

For a decade or so after its signature, the Treaty functioned with no major complaints from the rest of the world. Having garbed their territorial interest in impartial scientific interest and having invited non-claimants who had nonetheless been active during the International Geophysical Year, the original claimants and potential claimants deflected any immediate criticisms. Moreover, the fact that the Treaty's preamble underlined that this was done in the interest of the whole world was a way of expressing that, even if both the enfranchised and the representatives were a self-selected group, all humanity ought to feel represented, that is, as constituting the Antarctic demos.

However, this changed as soon as the prospect of turning Antarctica into a profitable economic enterprise began to be discussed. At that point, those who were only symbolically part of the Treaty (to wit, the big section of 'mankind' that was neither enfranchised nor empowered to make decisions) quickly pointed out its lack of representativeness and demanded actual participation.

4.3 The 'Question of Antarctica'

Until the early 1970s, the Antarctic Treaty could defend its political constitution as fair enough: those who had decision-making power were those investing in science on the continent without getting any immediate returns—not, at least, in terms of immediate, obvious material rewards. Instead, they were giving something back to all of humankind by way of knowledge, plus keeping the continent non-militarized and non-nuclearized. They had enfranchised themselves and appointed themselves as representatives of humanity in a place so remote and apparently so far from the political and economic radar that the arrangement was seen as basically harmless. From the outside, the active members of the Treaty could be seen as benefactors who had taken on their shoulders, self-willingly, the responsibility to manage a whole continent (representing no less than ten percent of the earth's total landmass) for the sake of all.

But everything changed when discussions about prospecting and mineral extraction in Antarctica began. This happened from the early 1970s, when a

[13] This was the case of Belgium, which practically disappeared from Antarctica after signing the Treaty, without this diminishing in any way its decision-making power within it: Alan D. Hemmings, 'Considerable Values in Antarctica', *The Polar Journal* 2, no. 1 (2012): 139–56, 141, http://www.tandfonline.com/doi/abs/10.1080/2154896X.2012.679565.

looming global energy crisis made the parties flirt with the idea that a solution might be found in the extraction of Antarctic minerals, especially gas and oil. Alarmed by the possibility that a privileged group of now 13 countries would benefit exclusively from the resources of a whole continent, outsiders started questioning the Treaty.[14]

In 1975, Hamilton Shirley Amerasinghe, Sri Lanka's ambassador to the UN and president of the Law of the Sea Conference, claimed that, in Antarctica

> *there are vast possibilities for a new initiative that would rebound to the benefit of all mankind. Antarctica is an area where the now widely accepted ideas and concepts relating to international economic cooperation with their special stress on the principle of equitable sharing of the world's resources, can find ample scope for application*[15]

Just like the High Seas, Amerasinghe suggested, Antarctica ought to become a site from which everyone could profit on an equal basis, not just a self-selected few. Four years later, the Peruvian diplomat Alvaro de Soto labelled the Antarctic Treaty as a form of apartheid that discriminated against African states especially.[16] While the wording was too strong, there was some truth to the claim if one looked at the composition of the Treaty at the time: out of the thirteen consultative parties, there was only one African state (South Africa), one Asian state (Japan), and two South American ones (Chile and Argentina).[17]

But the first sustained criticism of the Antarctic Treaty was lodged by the Malaysian prime minister, Muhammed bin Mahathir, at the United Nations General Assembly Meeting in 1982. In his speech, he labelled the Treaty as 'an agreement between a select group of countries [that] does not reflect the true feelings of the members of the United Nations or their just claims'.[18] The year was not random: Mahathir was speaking at a time when the then fourteen consultative parties had started negotiating a Convention for the Regulation of Mineral Resource Activities in Antarctica, or CRAMRA for short (hereinafter, the Mineral Convention).[19] Moreover, it was the final year of the Third United Nations Conference on the Law of the Sea, which—it was hoped—would turn the yet unclaimed wealth of the deep seabed into Common Heritage of

14 The thirteenth country to join as a consultative party was Poland, in 1977.

15 Deborah Shapley, *The Seventh Continent: Antarctica in a Resource Age* (Resources for the Future, 1985), 222.

16 Ibid., 230.

17 To this day, South Africa remains the only African state in the Antarctic Treaty.

18 United Nations General Assembly, 'Official Records, 29 September 1982', A/37/PV.10, 132.

19 West Germany became a consultative party in 1981.

Mankind, ensuring that its potential exploration and exploitation would benefit all states on the planet and not just those with financial and technological means.

For Mahathir, as for many other leaders of developing countries advocating for a more equitable New International Economic Order, all uninhabited areas in the world that were not yet under someone's undisputed sovereignty should be considered as belonging to the international community and should be administered (i.e. economically developed), especially for the benefit of the worst off. Antarctica, according to him, was the largest of these areas. Regardless of whether its administration fell on the United Nations or on the 'occupants' (*sic*) acting as 'trustees for the nations of the world', the goal was clear:

> *At present the exploitation of the resources of Antarctica is too costly and the technology is not yet available, but no doubt the day will come when Antarctica can provide the world with food and other resources for its development. It is only right that such exploitation should benefit the poor nations as much as the rich.*[20]

Mahathir became the most radical critic of the Antarctic Treaty, and his speech triggered what came to be known as the 'Question of Antarctica' at the United Nations, which ran at first annually, then biennially, and then triennially until 2005.[21] From then onwards, it was dropped from the agenda after the issue of economic exploitation was put aside and environmental protection became the value instead (more on this below).

When the remote prospect of benefiting economically from the exploitation of Antarctic mineral resources became probable, the question of who should decide how to proceed and who would be affected (here in the sense of benefited) by those decisions became relevant. What Mahathir and others began to dispute, in short, was not only the distribution of the benefit from the potential exploitation of Antarctica but also the constitution of the Antarctic demos and, especially, of its enfranchised and representatives. The critics of the Antarctic Treaty disputed the authority and legitimacy of the original founding group

[20] United Nations General Assembly, 'Official Records, 29 September 1982', 132.

[21] See Dodds, 'Post-Colonial Antarctica', 66–67. On the Question of Antarctica, see Peter J. Beck, 'The United Nations and Antarctica, 2002: "Constructive Engagement" Between the Antarctic Treaty System and the United Nations', *Polar Record* 39, no. 3 (2003): 231–38, https://doi.org/DOI: 10.1017/S0032247403003103; Peter J. Beck, 'Twenty Years On: The UN and the "Question of Antarctica", 1983–2003', *Polar Record* 40, no. 3 (2004): 205–12, https://doi.org/DOI: 10.1017/S0032247403003371; and Peter J. Beck, 'The United Nations and Antarctica, 2005: The End of the "Question of Antarctica"?', *Polar Record* 42, no. 3 (2006): 217–27, https://doi.org/10.1017/S003224740600533X.

in setting the governance rules for a whole continent.[22] They disputed the fact that joining the group required an economic and technological investment that was beyond their means, *de facto* preventing them from taking part (even if, on paper, the Antarctic Treaty was open for everyone to join). Finally, and most importantly, they disputed that the initial self-enfranchised group joined by two others who had met the criterion chosen by the self-enfranchised group (Poland and Germany) would now benefit economically from a continent over which—they claimed—no one ought to have more rights than any other.

Put in the terminology of democratic theory, Mahathir and others tacitly invoked the *All Affected Interests Principle*, to recall, that everyone who would be relevantly affected by the decisions of a political entity should be represented by that political entity—where being relevantly affected referred specifically to the prospects of economic gain.[23] There were, according to Mahathir, two ways to achieve this. On the one hand, the more demanding option was internationalization, by giving the governance of Antarctica to the United Nations, thus ensuring that the interests of all humankind, through their states' representatives, would be considered at the negotiating table. This was the path that India had proposed in the 1950s and that had been rejected by the Antarctic Treaty founders.[24] On the other hand, the less demanding option, as Mahathir himself conceded, was to keep the Antarctic Treaty as it was, but have the 'occupants acting as trustees'.[25] These trustees would be closely monitored and checked by the whole world to ensure that what they did affected everyone positively, or at least not negatively. Armed with a dose of political realism, the Malaysian prime minister realized that, absent an Antarctic democracy, the least bad option would be to fulfil what Robert E. Goodin has called a 'democratic third best', namely:

[22] These countries included Antigua and Barbuda, Bangladesh, Egypt, Bolivia, Malaysia, Nigeria, Pakistan, the Philippines, Sri Lanka, and Zambia.

[23] The rejection of a potential Mineral Convention by third parties to the treaty is a good example of why the *All Affected Interests Principle* seems more adequate than the *All Subjected Interests Principle*. If the Convention had been implemented, those subjected would have mainly been the citizens of the member states, even though the Convention would have had momentous consequences for those beyond them. If one argues for a narrow reading of the *All Subjected Interests Principle*, it seems unsatisfactory in this case. Alternatively, if one argues for an expansionary reading that includes members of third-party states as subjected, then *All Subjected* become subsumed under *All Affected*.

[24] Dodds, 'Post-Colonial Antarctica'; Sanjay Chaturvedi, 'Rise and Decline of Antarctica in Nehru's Geopolitical Vision: Challenges and Opportunities of the 1950s', *The Polar Journal* 3, no. 2 (2013): 301–15, https://doi.org/10.1080/2154896X.2013.868087. Antigua and Barbuda echoed this idea, suggesting that Antarctica should be governed by a secretariat housed at the UN, and that 'a system of licensing should be worked out with the benefits of [resource] exploitation being placed into a special fund from which all nations could draw'. This fund would be managed by the World Bank, giving grants to least developed states, soft loans to less developed states, and hard loans to the more developed: United Nations General Assembly, 'Official Records, 28 November 1984', A/39/PV.50.

[25] United Nations General Assembly, 'Official Records, 29 September 1982', 132.

> *the price of not enfranchising everyone we ideally should is that we would have to pay them off for any harms we inflict upon them and accede to their demands to fair recompense for any benefits we derive from the wrongfully disenfranchised.*[26]

In this case, states that were not Antarctic Treaty members (namely, most of the world at the time) should at least be economically compensated if not invited to join the decision-making table.

It is important to remember that the Question of Antarctica came about as part of a more general struggle for global distributive justice. It was the time when the idea of a New International Economic Order was being pushed by developing states, many of them recently rid of their colonial ties. As with the deep seabed, Antarctica was seen as a potential treasure trove, the equitable exploitation of which would allow these countries to 'grow', develop, and catch up with the others. That was the motivation to label Antarctica as 'Common Heritage of Mankind': not to protect it for future generations or for its own sake, but rather to make it available for everyone who might profit from its exploitation.[27]

As with other demands for a New International Economic Order, the demands to make the Antarctic Treaty less exclusive and more democratic did not succeed. Peter Beck has noted that having left ambiguous what counted as 'substantial scientific research' was an advantage for the Treaty members, because it allowed 'the interposition of discretionary, pragmatic and policy considerations'.[28] In this specific context, it meant that the parties displayed an extra dose of generosity when interpreting it and relaxed the initially high bar for acceptance as full member. Poland and West Germany, who became consultative parties in 1977 and 1981, respectively, conducted a series of marine and geological research expeditions, and established permanent stations, both

[26] Goodin, 'Enfranchising All Affected Interests', 68.

[27] There is nothing in the concept of Common Heritage of Mankind that necessarily links it to exploitation, but this is how it was conceived of at the time. The 'preservationist' interpretation came much later. Christopher C. Joyner, *Governing the Frozen Commons: The Antarctic Regime and Environmental Protection* (University of South Carolina Press, 1998), 33–34. See also Surabhi Ranganathan, 'Ocean Floor Grab: International Law and the Making of an Extractive Imaginary', *European Journal of International Law* 30, no. 2 (2019): 573–600, 594, https://doi.org/10.1093/ejil/chz027. Ranganathan describes the Deep Seabed Regime as treating the deep seabed essentially as a mining site.

[28] Peter J. Beck, *The International Politics of Antarctica* (Croom Helm, 1986), 194. See also Orrego Vicuña's circular defence of the criterion: 'It is certainly not the case that any research whatsoever is sufficient reason for granting the special responsibility, but rather only those research efforts whose significance clearly demonstrates the presence of the necessary qualifications': Francisco Orrego Vicuña, 'Antarctic Resources Policy: An Introduction', in *Antarctic Resources Policy: Scientific, Legal and Political Issues*, ed. Francisco Orrego Vicuña (Cambridge University Press, 1983), 3.

at a high economic cost.[29] Instead, for Brazil and India, who joined in 1983, the bar was set much lower: they were both accepted after having conducted a couple of expeditions and promising that they would implement an Antarctic research program.[30] China also joined in 1983, helping to balance the presence of developed and developing countries and assuaging the critique of elitism. Moreover, the Antarctic Treaty Meetings, criticized for their secrecy, became open to non-consultative parties from that year onwards (the most sensitive meetings, however, continued behind closed doors).[31] All along, the idea of democratizing Antarctic governance remained a non-starter for the original members, especially the claimants. The self-appointed trustees appeased their wards instead by promising that they would continue to fulfil what they had committed themselves to do, namely, to act 'in the interest of all mankind'.[32] They made an effort, that is, to show that Mahathir's less demanding interpretation of the *All Affected Interests Principle* could be harmonized with the Treaty as it was. To be credible, however, the prospect of making economic gains had to be given up.

4.4 Freezing interests

If all those potentially affected by the decisions of a political entity should somehow take part in the decision-making process, but consulting all those potentially affected proves undesirable, then one can solve the problem by avoiding affecting them altogether, even if this also means avoiding affecting oneself. This is one way to explain the abandonment of the Mineral Convention in the late 1980s and its replacement by a wholly different instrument: the Protocol on Environmental Protection to the Antarctic Treaty (the Environmental Protocol), signed in 1991 and ratified in 1998.

As was shown, from the mid-1970s through the 1980s, developing countries led by Malaysia realized that the politics of Antarctica could affect them by leaving them out of Antarctic decision-making and therefore also out of the potential exploitation of a continent that they thought ought to belong to all humanity. If the Antarctic Treaty parties wanted to proceed with the division,

[29] For Poland, this meant spending around three million USD in the 1970s, while for West Germany the investment was 110 million USD worth of equipment for the Alfred Wegener Institute, founded in 1980 (Beck, *International Politics of Antarctica*, 200).
[30] Ibid.
[31] Shapley, *Seventh Continent*, 220.
[32] Conference on Antarctica, 'The Antarctic Treaty', Preamble.

then it was only fair that all states should participate in the decision-making process or at least benefit from an economic perspective. Instead, what happened was that the consultative parties chose not to let anyone benefit in that sense, therefore not 'affecting' anyone. That way, things could continue as they were, and there was no reason for anyone to hold a grudge. Some have even suggested that giving up the Mineral Convention came as a relief, especially for the claimant parties. The issue of how the frozen territorial claims could be sustained under such a convention had been a matter of constant worry since the idea began to be discussed. Some even speculated that, depending on how the convention was framed, 'sovereignty over territory . . . in the Antarctic would be reduced to a mere vestige of rights'.[33] It is no coincidence, then, that it was two of the original claimants, Australia and France, who decided to change their stances from approval to opposition.[34]

A less cynical explanation of why the Environmental Protocol replaced the Mineral Convention is that the mood was in the air. On January 28, 1989, the ship *Bahía Paraíso* ran aground near the US Palmer Station and began leaking oil into the sea, producing the worst environmental disaster in Antarctica to date.[35] A few months later, on 24 March 1989, the oil tanker *Exxon Valdez* ran aground in Prince William Sound, in Northern Canada, spilling 11 million gallons or more of crude oil into the area. Adrian Howkins persuasively suggests that these environmental catastrophes came as a warning and as a signal of the fragility and vulnerability of the polar regions. Together with Greenpeace's campaign to turn Antarctica into a World Park, and with a growing global awareness of Antarctica's importance for the planet's 'health', these gave the political momentum needed for the Environmental Protocol.[36] By virtue of its Article 7, any activity relating to mineral resources, except for scientific research, was banned, and all other activities in Antarctica became subject to environmental considerations.

Through the Protocol, the consultative parties gave greater force to the idea that the best suited to manage Antarctic affairs were 'those who knew'—where knowledge was understood as scientific knowledge, and where it was assumed that scientific knowledge and environmental protection went hand in hand. It

[33] F. M. Auburn, 'Antarctic Minerals and the Third World', *FRAM: The Journal of Polar Studies* 1 (1984): 200–23, 208.

[34] Australian politicians in particular were worried that signing the Mineral Convention would mean losing sovereignty over their Antarctic territory: Doaa Abdel-Motaal, *Antarctica: The Battle for the Seventh Continent* (Praeger, 2016), 108–9.

[35] National Oceanic and Atmospheric Administration (NOAA), 'M/V Bahia Paraiso; Palmer Station, Antarctica', IncidentNews, 28 January 1989, https://incidentnews.noaa.gov/incident/6658#!.

[36] Howkins, *Polar Regions*, 162ff.

was as if the interpretation of what counted as 'affectedness' by the *All Affected Interests Principle* had been switched from economic to environmental. While the availability of energy sources was still an issue, by the end of the 1980s the negative effects that humans were having on the planet became evident and alarming. The Environmental Protocol marked a new era, where being relevantly affected by Antarctic governance no longer meant standing to gain or lose from economic exploitation, but rather standing to gain from its environmental preservation and standing to lose from its neglect. Environmental preservation was thought to safeguard Antarctica's key function as a global thermostat, a regulator of atmospheric and oceanic circulation, and a freshwater reserve which, in case of melting, would dramatically impact sea levels around the world. This reasoning appears explicitly in the initial proposal for the Protocol, presented by Australia and France:

> *The Antarctic region, including the Southern Ocean, is inextricably linked with the global environment and thus has an essential role in studies of it. Interactions between the atmosphere, ice, ocean and living systems affect the entire global system through mechanisms still insufficiently understood. Antarctica has a major influence on the global environment. The Antarctic ice sheet contains 70% of the world's fresh water. Any significant changes to the Antarctic environment could have global consequences on atmospheric and oceanic circulation and upon sea levels.*[37]

To better understand the role of Antarctica in the planet's systems, furthermore, promoting science as the main activity and keeping the continent as the most pristine open-air laboratory in the world became more important than ever. That science remained as the Treaty's political currency thus looked more justifiable, too.

The Environmental Protocol arguably became the most demanding international legal instrument for environmental protection worldwide, ensuring the preservation of a whole continent for that end. Under its framework, the Antarctic demos remained humanity, whose best interests required protecting the continent, a protection that would be carried out by those states that performed science in it—and where scientific results would purportedly reinforce the idea that protection was important.

[37] Australia and France, 'A Joint Australia/French Proposal in the Form of a Paper Including a Draft Recommendation for ATCM XV: Comprehensive Measures for the Protection of the Antarctic Environment and Its Dependent and Associated Ecosystems', XV ATCM/WP2, Paris, October 9–19, 1989, 1, https://documents.ats.aq/ATCM15/wp/ATCM15_wp002_e.pdf.

One could say that, thanks to the Environmental Protocol, the Antarctic Treaty succeeded in appeasing its critics (even Malaysia ended up joining as a non-consultative party in 1998) and retained its legitimacy as an international body. In 2005, the Question of Antarctica was discussed for the last time at the United Nations. In the meantime, some non-consultative parties gained full status while new members joined.[38] From its beginning with twelve signatories, today the Treaty counts fifty-eight members in total, twenty-nine of them consultative and twenty-nine of them non-consultative. Even though it includes fewer than a third of all existing states, its members account for two-thirds of the global population (it is still skewed towards the 'Global North', though, with only one African member and just a few from Asia and Oceania).

As I show next, the fact that political power is the privilege of those who can afford the performance of science on the continent remains one of the main critiques of the Treaty, earning it its characterization as a colonial device. At the same time, I suggest that, regardless of the intentions of individual states, the Treaty post–Environmental Protocol achieved something remarkable: namely, collective restraint. Thanks to it, Antarctica remains the only place on earth where humans have set different rules of engagement (or, rather, non-engagement) with nature; rules that take not only our interests into account, but potentially those of non-human others.

4.5 The Antarctic Treaty as a colonial device?

The only time I have been to an Antarctic Treaty Meeting was in 2016, in Santiago de Chile. I was a visitor at the Chilean Antarctic Institute in Punta Arenas, and I participated as a member of the Chilean delegation (individuals as such cannot take part in the meetings unless they join a national delegation or some observer or expert body). The big conference room, in a five-star hotel, was arranged so that the heads of delegation of the consultative parties sat around the inner circle, surrounded by those from the non-consultative parties, together with other representatives from observer and expert institutions, such as the Scientific Committee on Antarctic Research, the Convention for the Conservation of Antarctic Marine Living Resources, the International Meteorological Organization, the International Association of Antarctic Tour Operators, and the environmental NGO Antarctic and Southern Ocean Coalition. To my surprise, many seats from non-consultative

[38] Beck, 'United Nations and Antarctica, 2005'.

parties were empty. The explanation I got made sense: governments had priorities and coming all the way to Chile for 10 days to listen to what others would end up deciding about Antarctic governance was not one of them. Looking for a good vantage point in the middle of the room, I sat halfway through the alphabet, in the place reserved for the Mongolian delegation. I was not alone: attendees from other national delegations were all around me, using the vacant seats of Costa Rica, Cuba, Denmark, Greece, Iceland, Papua New Guinea, and Slovenia, among others.[39]

This personal experience corroborates what many authors have criticized as the chronic elitism of the system and its *de facto* closed nature. The fact that science is the most important activity to be pursued in Antarctica has led to its frequent description as 'the continent for science',[40] but the fact that science was, and still is, the 'political currency' in the Treaty has also come under intense scrutiny. In Aant Elzinga's words: 'Science is the instrument needed to exercise power and influence on its arena, to the exclusion of those that cannot or have decided not to mobilize this resource in the way that the Antarctic Treaty prescribes'.[41] Under the Antarctic Treaty, Antarctica becomes a continent not just *for science,* but also *by science*, where production of science (symbolic capital) is a prerequisite criterion for new countries to become [full-blown] members. Underlying this is the notion that 'science in Antarctica is in certain important respects *a continuation of politics by other means;* concomitantly, *politics is sublimated in scientific activities*'.[42] As historian Peder Roberts summarizes, 'Antarctic science has always been a geopolitical performance'.[43]

The tight connection between the performance of science on the continent and the political motives of the performers is a recurrent topic of analysis among Antarctic scholars.[44] An example is the correlation between the level of

[39] That year, only eleven out of twenty-four non-consultative parties attended the meeting: Secretariat of the Antarctic Treaty, 'Final Report of the Thirty-Ninth Antarctic Treaty Consultative Meeting (Santiago de Chile, 23 May–1 June 2016)', vols. 1 & 2, Buenos Aires, 2016, 281–87, https://documents.ats.aq/ATCM39/fr/ATCM39_fr001_e.pdf and https://documents.ats.aq/ATCM39/fr/ATCM39_fr002_e.pdf. This low level of attendance is the rule rather than the exception.

[40] R. S. Lewis, *A Continent for Science: The Antarctic Adventure* (Secker & Warburg, 1965).

[41] Aant Elzinga, 'Geopolitics, Science and Internationalism During and After IGY', in *Boletín Antártico Chileno: Second SCAR Workshop on the History of Antarctic Research (Santiago, Chile, 21 to 22 September 2006)*, ed. Reiner Canales Cabezas (La Prensa Austral, 2006), 74, www.inach.cl/wp-content/uploads/2009/10/boletin-historico-agosto-2009.pdf.

[42] Ibid., 71, his emphasis. See also Aant Elzinga, 'Antarctica: The Construction of a Continent by and for Science', in *Denationalizing Science*, ed. Elisabeth Crawford, Terry Shinn, and R. D. Whitley (Springer, 1993), https://doi.org/10.1007/978-94-017-1221-7_3 and Klaus Dodds, 'Assault on the Unknown: Geopolitics, Antarctic Science and the International Geophysical Year (1957–8)', in *New Spaces of Exploration: Geographies of Discovery in the Twentieth Century*, ed. Simon Naylor and James R. Ryan (I. B. Tauris, 2015).

[43] Peder Roberts, *The European Antarctic: Science and Strategy in Scandinavia and the British Empire* (Palgrave Macmillan, 2011), 157.

[44] See, for example, Brady, *China*, chapter 5.

political involvement in the Treaty and the level of scientific production: the seven claimants plus the United States and Russia are the leading countries in terms of scientific output.[45] Despite the Treaty chanting the wonders of international scientific cooperation and collaboration, moreover, science in Antarctica remains an eminently national affair. Jacques-Yves Cousteau's dream of a truly international research station run by the Scientific Committee on Antarctic Research (SCAR) remains just that, more than 30 years after: a dream.[46] At the same time, the criteria for what counts as 'substantial scientific research' have kept evolving. While the establishment of a station was a *sine qua non* until 1990, that year the Netherlands was the first to become a consultative party while remaining homeless on the continent. Doing good science in Antarctica, the Dutch argued, did not require establishing physical presence in the form of a building. In fact, doing science without leaving a trace was the way forward. More recently, collaboration with ongoing programmes and use of already established scientific infrastructure are encouraged, and the way to measure the quality of the research published has also become more demanding.[47] Despite this evolution, however, the principle of dividing countries into two tiers depending on their scientific performance has persisted since the Treaty's inception.

In the last chapter, I suggested that the territorial claims in Antarctica were an expression of colonialism and thus morally problematic. Here, I address the contention that the Antarctic Treaty itself is also a colonial device in two main ways. Firstly, by virtue of the Treaty, the application of the colonial strategies of continuity, contiguity, and the sector principle in Antarctica, and even the legitimacy of an eminently colonial treaty like Tordesillas, remained unchallenged. They were frozen with the territorial claims and preserved up to this day, without the slightest sign of the claimants reflecting on their problematic nature. This is what leads Klaus Dodds to affirm that the Antarctic Treaty ended

[45] John R. Dudeney and David W. H. Walton, 'Leadership in Politics and Science within the Antarctic Treaty', *Polar Research* 31 (2012): 1–9, https://doi.org/10.3402/polar.v31i0.11075.

[46] On Cousteau's ideal for Antarctica, see Janet Martin-Nielsen, *A Few Acres of Ice: Environment, Sovereignty and Grandeur in the French Antarctic* (Cornell University Press, 2023), 166–68, https://doi.org/10.7591/cornell/9781501772092.001.0001. On the persistent nationalism of Antarctic programmes, see Alan Hemmings, Sanjay Chaturvedi, Elizabeth Leane, Daniela Liggett, and Juan Francisco Salazar, 'Nationalism in Today's Antarctic', *The Yearbook of Polar Law* VII (2015): 531–55, https://doi.org/10.1163/2211-6427_020.

[47] In 2017, the Antarctic Treaty adopted new rules on how to interpret the science criterion. These rules include evidence of scientific quality measured by 'details of citations of relevant papers that scored well in a science citation index'; 'data cited in publications that score well in a science citation index and . . . data contributed to Antarctic scientific programmes and databases'; and 'examples of research prizes or formal recognition of accomplishments': Secretariat of the Antarctic Treaty, 'Guidelines on the Procedure to Be Followed with Respect to Consultative Party Status (Decision 2, ATCM XL—CEP XX, Beijing)', adopted 1 June 2017 https://www.ats.aq/devAS/Meetings/Measure/653.

up 'rewarding' colonial occupation and annexation,[48] and Alan Hemmings to observe that 'the present Antarctic dispensation . . . was arrived at . . . as a result of an imperial and colonial model now generally repudiated'.[49] Instead of giving up their claims, in short, the original claimants turned them into a cornerstone of the Treaty. As mentioned before, when the Question of Antarctica became a topic for discussion at the United Nations General Assembly at the beginning of the 1980s, this fact was underlined by non-aligned countries who wished for Antarctica to be considered Common Heritage of Mankind and denounced the claims as belonging to a bygone era. As the Pakistani delegate summarized, the purported rights of the claimant states were based on territorial claims that were made during the colonial era, but 'in the present-day world, when the process of decolonization has almost been completed, the colonial premise on which these claims were based has been rejected'.[50]

Secondly, the way in which the Treaty was constituted and how its original members were chosen resonates with what Lea Ypi characterizes as the central wrong of colonialism; namely, the political domination of one group over another, where the domination involves 'the creation and upholding of a political association that denies its members equal and reciprocal terms of co-operation'.[51] In a twist to Ypi's account, the wrong here did not consist of the constitution of a political association that was not negotiated under equal and reciprocal terms *by its members*, but in the treatment of many of those *who ought to have been members* in the initial establishment of the association as wards to be represented, and in their ongoing status as such. In constituting the Treaty, the twelve original signatories (three-quarters of whom were claimants and potential claimants) appointed themselves not only as guardians of Antarctica, but also as guardians—in the sense of legal representatives—of the rest of humanity for Antarctica. And they picked a criterion (science) to ensure that only states that were developed enough would be able to take an active role in its politics.[52] As Yelena Yermakova puts it, using a power-driven-path-dependence approach, the initial power imbalance was reinforced and locked in through institutionalization.[53]

[48] Dodds, 'Post-Colonial Antarctica', 63.

[49] Hemmings, 'Security Beyond Claims', 77.

[50] United Nations General Assembly, 'Official Records, 28 November 1984'.

[51] Ypi, 'What's Wrong', 158. Notice that in this account colonialism belongs in a larger class of wrongful relations of domination between people.

[52] See Elzinga, 'Construction of a Continent' and Yelena Yermakova, 'Legitimacy of the Antarctic Treaty System: Is It Time for a Reform?', *The Polar Journal* (2021): 1–18, https://doi.org/10.1080/2154896X.2021.1977048.

[53] Yelena Yermakova, 'Domination as a Legacy of the Colonial Origins and Structure of the Antarctic Treaty System', in *Colonialism and Antarctica: Attitudes, Logics, and Practices*, ed. Peder Roberts

I count myself among those who have been critical to these aspects of the Treaty and have suggested the need for reform.[54] For one thing, the territorial claims, as relics of a past era, should be recognized as such and be given up. To stay with the architecture metaphor, one should see them as part of the initial scaffolding rather than as cornerstones of the Treaty: maybe indispensable to set it up but no longer needed once the building is in place. For another thing, the locked-in hierarchy among states should be revisited, and all those whose interests are potentially affected by Antarctic governance ideally ought to take part. Here, however, it is important to avoid perpetuating the methodological nationalism and anthropocentrism that have operated as the default. Extending membership only to states and only for the sake of human interests, in other words, would be insufficient and inadequate. As I suggest in the coming chapters, there is a case for representing nature in Antarctic governance, and there is a case for this to be done not through the usual representatives that have operated so far (namely, states), but through others. I grant that this would require a significant leap of political will and imagination, and that the mechanisms for how to do this would have to be created from scratch. However, there is a precedent for creative political imagination in Antarctic governance which gives hope.

Despite the criticism, thus, I think it is still possible to hail the results achieved by the Antarctic Treaty as impressive. Thanks first to the Treaty and then to the Protocol, Antarctica has remained for over six decades as a continent where states have shown self-restraint. An exclusive dedication to peace, scientific research, and the protection of the environment have turned it into a thing of its own; a place where, even against their initial motivations, the states involved ended up creating a unique model of territorial governance, focused on *protecting* rather than *exploiting*. Those fond of geopolitics and international relations may object that it is just a matter of time before the continent fully enters global value chains and loses its status as 'the last wilderness'. For them, it is inevitable that Antarctica will be exploited just like any other site, and that its resources will end up divided among the most powerful.[55]

My reply to this (which I expand in Chapter 9) is that, even if that were the case, these past 60 years would remain an example of territorial governance for

and Alejandra Mancilla (Manchester University Press, 2024), https://doi.org/10.7765/9781526170644.00018.

[54] Mancilla, 'Decolonising Antarctica'.

[55] See, for example, how some political scientists have developed a scenario of future militarization for Antarctica based on classical geopolitical analysis: Jeffrey McGee, David Edmiston, and Marcus G. Haward, *The Future of Antarctica: Scenarios from Classical Geopolitics* (Springer, 2022), 159.

other places where a focus on guardianship and tighter coordination between states could be crucial for their successful environmental protection. It would also be a reminder that international law is our own creation, and not something set in stone. It is up to us to devise new frameworks to address new challenges, just like the Antarctic Treaty did.

4.6 Concluding remarks

In this chapter, I examined the construction of the Antarctic demos and its representatives, looked at the main criticisms against it, and concluded that, despite them, the Treaty has achieved something remarkable: namely, leaving a whole continent for peace, science, and the preservation of the environment, and creating a model of territorial governance that may serve as inspiration beyond Antarctica. By interpreting the 'interests' in the *All Affected Interests Principle* as environmental rather than economic, from the 1990s onwards the Treaty justified its structure to its non-members in the following terms: insofar as all humanity had a stake in the environmental protection of Antarctica, the Treaty members would remain focused on this goal.

In conclusion, I suggest that the Environmental Protocol has the tools for extending the Antarctic demos and its representatives even further, although those tools have not yet been used for this purpose. Using them, however, would require becoming mindful of methodological nationalism and methodological anthropocentrism.

As critics of methodological nationalism have pointed out, this cognitive bias leads to the reification of state-centred conceptions of territory and space, impeding us from seeing or conceiving of other possible modes of territorial organization and relations.[56] In Antarctic governance, methodological nationalism lurks in the background: at no point during the discussions leading to the Treaty or, later, to the Environmental Protocol, was it suggested that Antarctica ought to be governed by a different type of authority altogether, neither *national* nor inter-*national*. It was assumed that nation-states should be the main actors carrying the process forward and representing the interests of their constituents and of human beings in general. Furthermore, there was no talk of including other political entities in the process, such as non-governmental organizations and other groups. Even though this place was unique not just from a physical and geographical perspective, but also from

[56] Sager, 'Political Philosophy'.

a political perspective—the only continent with no human populations!—it was not even thought that maybe states ought to stay out of it altogether, as philosopher Holmes Rolston III suggested decades later.[57] Antarctic scholars who deny the 'Antarctic exceptionalism thesis' point out that the history of Antarctica is not something of its own, but stands in continuity with global history.[58] The fact that methodological nationalism was ingrained in the politics of the continent from the very beginning adds force to their point: a place that was treated as an exception to the usual rules in some aspects was not treated as an exception when it came to deciding who the main actors would be.

Secondly, there is methodological anthropocentrism. The discussion so far has uncritically assumed that human interests are the only ones that matter in themselves and, thus, that human interests are the only ones we should care about in political deliberation and law creation. This has prevented the parties from considering the possibility that other beings may also be political and legal subjects. This is the case even though some of the legal documents they have agreed upon leave the door open for the political and legal consideration of non-human beings and entities. Moreover, the criticism against the Treaty focused on the *scope* of its demos but never questioned the *kinds* of entities that ought to constitute it in the first place. It was assumed that it was humans and humans only who had a stake in the continent—never mind the fact that it was the only place on earth with no native humans and no permanent human settlements. Those fighting for wider representation as much as those who had established the rules thus agreed that the question was *how* to delimit the human demos and choose its representatives in an 'empty' territory, rather than *who* the demos ought to be. The fact that the territory was home to a 'magnificent, wild, self-regulating and awe-inspiring living community of inter-related ecological beings'[59] did not make any difference—or, if it did, it was only when these were considered not as subjects, but as resources.

One could object that it is unfair to judge real actors in the real political arena by standards used to evaluate the role of theorists engaged in scholarly analysis. Moreover, most of these actions happened at a time when methodological nationalism and anthropocentrism were mostly taken for granted

[57] Holmes Rolston III, 'Environmental Ethics in Antarctica', *Environmental Ethics* 24, no. 2 (2002): 115–34.

[58] See, for example, Alan D. Hemmings, 'The Philosophy of Law in the Antarctic', in *Philosophies of Polar Law*, ed. Dawid Bunikowski and Alan D. Hemmings (Routledge, 2020), 13–29.

[59] Cormac Cullinan, 'The Antarctic Realm: Exploring the Potential of Antarctica as a Sovereign Entity', presentation at the joint workshop 'Antarctica and Rights of Nature: Historical, Moral and Legal Perspectives', organized by the projects 'Greening the Poles' and 'Dynamic Territory' (Lysebu, August 2022).

even in academia: why expect the actual national delegates and international lawyers to engage in that kind of self-reflection (which was not even necessarily compatible with the furthering of their own interests)? My response is that it is sometimes the actors on the ground who come up with new perspectives and critiques that are then systematized by theorists. As will be seen in Chapter 6, the demands for a radically different Antarctic demos have not come so much from academics as from activists and environmental NGOs. Thus, even though it might be true that, in the first five decades of the Antarctic Treaty, methodological nationalism and anthropocentrism have been the default starting points, this does not mean that actors must necessarily be blind to them and uncritical of the framework within which they are assumed to ask their questions.

In the next two chapters, I suggest that the tools for thinking differently about the Antarctic demos and its representatives are already present in the Environmental Protocol. If Antarctica has intrinsic value, as this Protocol claims, and if this intrinsic value makes its possessors worthy of political and legal consideration and maybe also representation, then not just humans but also non-human animals and even Antarctica as a whole should constitute the Antarctic demos and be represented.

5

Representing Sentient Antarctica

One of the most discussed passages in the Environmental Protocol affirms that

The protection of the Antarctic environment and dependent and associated ecosystems and the intrinsic value of Antarctica, including its wilderness and aesthetic values and its value as an area for the conduct of scientific research, in particular research essential to understanding the global environment, shall be fundamental considerations in the planning and conduct of all activities in the Antarctic Treaty area.[1]

Rare in international law documents, this is a recognition that the place to be protected has not only instrumental value for humans but intrinsic value, and that the protection of this value will be a 'fundamental consideration' when determining what may and may not be done south of 60 degrees south.[2] This article reflected what some delegates had proposed as a necessary change of mindset in Antarctic governance during the discussions leading to the Protocol. As the head of the New Zealand delegation, Frank Wong, puts it in the first of four special meetings:

Interest in Antarctica's resources continues, inevitably, in a resource hungry world. At the same time people are demanding that we change our behaviour; that we become more conservation conscious; that we set aside areas from resource development; that we recognize that in some special places other values

[1] Antarctic Treaty System, 'Protocol on Environmental Protection (1991)', Article 3.

[2] The only precedent for the recognition of intrinsic value in an international law document is the Bern Convention, which claims 'that wild flora and fauna constitute a natural heritage of aesthetic, scientific, cultural, recreational, economic and intrinsic value that needs to be preserved and handed on to future generations': Council of Europe, 'Convention on the Conservation of European Wildlife and Natural Habitats (ETS no. 104)', 19 September 1979, https://rm.coe.int/1680078aff. Another example is the Convention on Biological Diversity, which states in its preamble that the parties are 'Conscious of the intrinsic value of biological diversity and of the ecological, genetic, social, economic, scientific, educational, cultural, recreational and aesthetic values of biological diversity and its components': United Nations Environment Programme (Secretariat of the Convention on Biological Diversity), 'Convention on Biological Diversity', entered in force 29 December 1993, published 2011, https://www.cbd.int/doc/legal/cbd-en.pdf.

From Sovereignty to Guardianship. Alejandra Mancilla, Oxford University Press.
DOI: 10.1093/9780197902769.003.0005

must be given primacy. The Antarctic continent is quintessentially one such place.[3]

The recognition that Antarctica has intrinsic value may be seen as a plunge into deep metaphysical waters and raises a number of questions: *Who* or *what* is the 'Antarctica' possessing value? On what grounds is that value assigned? And what is due to that entity or entities given that value? These questions would have been at the centre of the discussion leading to the Protocol and to its consequent implementation, had there been any philosophers in the room. In their absence, however, the international lawyers, diplomats, and national delegates who were present had no trouble allowing the expression 'intrinsic value of Antarctica' into the text without bothering to specify it. As is habitual in international law, a basic term of the agreement was left vague and open to interpretation.

Since the Environmental Protocol entered into force in 1998, it has been deemed sufficient to protect Antarctica without the need to argue for the continent having its own representative or guardian. However, this is not obvious. First, there is a growing movement in domestic and international environmental law advocating for the rights of nature. Since Christopher Stone's seminal article proposing that trees (or, rather, forests) ought to have legal standing, rivers and natural areas have been granted rights of their own, and nature has been recognized as a legal subject in national constitutions and laws.[4] Indigenous thought has had a key role in these developments: in places like Ecuador and New Zealand, assigning rights to nature has been a way of harmonizing Indigenous worldviews with the Western legal paradigm.[5] Second, despite being celebrated as creating the most protected environment on earth, one could argue that, more than 30 years after its signature, the Environmental Protocol is insufficient and inadequate to ensure the real protection of the continent. It is insufficient because it leaves out under-regulated and increasingly problematic activities (like tourism); further, it leaves another activity totally

[3] Republic of Chile, 'Interim Report of the Eleventh Antarctic Treaty Special Consultative Meeting', Viña del Mar, Chile, 19 November–6 December 1990, 57, https://documents.ats.aq/SATCM11_1/fr/SATCM11_1_fr001_e.pdf.

[4] See Christopher Stone, 'Should Trees Have Standing? Toward Legal Rights for Natural Objects', *Southern California Law Review* 45, no. 450 (1972): 450–501.

[5] See, for example, *Te Urewera Act 2014*, New Zealand Public Act no. 51, date of assent 27 July 2014, version of 5 April 2025, https://www.legislation.govt.nz/act/public/2014/0051/latest/whole.html; *Te Awa Tupua (Whanganui River Claims Settlement) Act 2017*, New Zealand Public Act no. 7, date of assent 20 March 2017, version of 27 August 2025, https://www.legislation.govt.nz/act/public/2017/0007/latest/whole.html; and *Constitución de la República del Ecuador 2008*, last modified 13 July 2011, https://www.oas.org/juridico/pdfs/mesicic4_ecu_const.pdf.

out of its realm, even though it is arguably the activity that could have the greatest impact on the Antarctic ecosystem—namely, fishing.[6] It is inadequate because there is a 'problem of fit' between the goals it sets for itself and the area where it is supposed to achieve those goals. Writing about environmental governance institutions, Oran Young defines 'fit' as 'a matter of (in)congruity between properties of the relevant ecosystems and attributes of the institutions created to guide human interactions with these biophysical systems'.[7] As I show in Chapter 8, the paradox at the heart of the Environmental Protocol is that, by limiting itself to actions south of 60 degrees south, the goal of protecting Antarctica, especially in a time of fast-paced climate change, becomes unachievable.

With the above in mind, one could propose that Antarctica should be represented in the Antarctic Treaty and maybe also in other global fora. If this is the way to go, then we need to answer the questions posed above: Who or what exactly is that 'Antarctica' in need of representation? What value is being protected? And, how to act accordingly? In this chapter and the next, I offer two lines of response and show their challenges, but also their promise. On the one hand, many animal ethicists and animal political theorists agree that sentience—that is, the capacity for feeling pleasure and pain—is the necessary and sufficient condition for 'being an object of primary moral concern'.[8] Sentience has intrinsic value, in this sense, because it is 'the ticket that admits something to the moral community'.[9] Starting from this, one could argue that the intrinsic value of Antarctica lies in its sentient inhabitants, and that protecting this intrinsic value requires granting them political and legal representation. This connection between moral standing and the right to be politically represented is not conceptually necessary, but it may be pragmatically advisable if one looks at the alternatives. On the other hand, inspired by the rights of nature movement and non-Western worldviews, one could argue that it is

[6] See, respectively, Bastmeijer et al., 'Regulating Antarctic Tourism' and Lynda Goldsworthy, 'Consensus Decision-Making in CCAMLR: Achilles' Heel or Fundamental to Its Success?', *International Environmental Agreements: Politics, Law and Economics* 22, no. 3 (2022): 411–37, https://doi.org/10.1007/s10784-021-09561-4.

[7] Young, 'Environmental Governance', 378.

[8] Dale Jamieson, *Ethics and the Environment: An Introduction* (Cambridge University Press, 2008), 70.

[9] Ibid. Theories about intrinsic value and the ways of relating intrinsic value to cognate terms like *inherent worth*, *final value*, and *ultimate value* are as numerous as there are philosophers writing about them. For the purposes of the discussion, I find Dale Jamieson's classification of intrinsic value into four main senses especially clear and straightforward: as a necessary and sufficient condition for being morally considerable; as opposed to instrumental, that is, as valuable for its own sake; as a non-relational property that inheres in some objects; and as value that is independent of valuers for its existence: Jamieson, *Ethics and the Environment*, 69–72.

Antarctica considered as one unit that should be represented. In this approach, it is the very existence of this wild community of life that is worth preserving. Antarctica has intrinsic value in the sense of being valuable for its own sake.[10]

In the next chapter I address the latter position. In what follows, instead, I focus on the first kind of response and ask whether sentient Antarctic animals should be politically represented and, if so, how. I first refer to their place in Antarctic history as mere resources, and then to their protection in the face of near extinction—where the consideration was still to preserve them as resources for future human use. I then explore what it would mean to locate the intrinsic value of Antarctica in its sentient beings for the purposes of political representation: should their interests be 'internalized' or 'encapsulated' by those already making decisions, or should they have their own representatives at the decision-making table? While favouring independent representation, I suggest that, regardless of the model that we pick, the underlying challenge is to determine which animal interests are most relevant. I suggest that, if we prioritize the interests of individual animals in not suffering and being harmed, we will get very different prescriptions than if we prioritize their interests in remaining as part of an autonomous collective and not interfered with by humans. I conclude that this gives us reason to think that a representational structure for non-human animals might need to be more diverse than those advocating it have granted so far. Furthermore, it gives us reason to think that extending membership in the Antarctic political community might not be enough, and that maybe Antarctica as a whole should be considered.

5.1 Sentient animals as resources

At length did cross an Albatross,
Through the fog it came;
As if it had been a Christian soul
We hailed it in God's name.

Samuel Taylor Coleridge, The Rhyme of the Ancient Mariner[11]

In *The Rhyme of the Ancient Mariner*, from 1834, the Romantic poet Samuel Taylor Coleridge tells the story of a ship caught by storms and brought all the

[10] Jamieson, *Ethics and the Environment*, 71. To put it in Kantian terminology, one could say that in both these senses of intrinsic value what matters is that its possessor should not be treated merely as a *means* but also as an *end*.

[11] Samuel Taylor Coleridge, *Selected Poetry and Prose* (Holt, Rinehart and Winston, 1951), 51.

way down to the Southern Ocean, where it encounters ice and nothing more until the Antarctic bird shows up. The mariner becomes acquainted with the albatross as days go by. As is usual with albatrosses, he keeps the ship company as it sails forth. All goes well until the mariner does not like the bird's presence anymore and shoots him. Too late, the mariner realizes that the bird was a luck charm. The wind stops, the ship is stuck in the middle of the ocean, and the crew recriminates him for his bad decision. The mariner regrets having killed the animal that was going to guarantee them a safe sailing forward.

To avoid spoilers, I will say nothing more about the story, but this much is enough to make my point; namely, that the thought did not cross even a sensitive soul like Coleridge's that killing the magnificent bird was not just wrong for its effects on humans, but a wrong to the bird. This instrumental rationale has informed Antarctic animal protection with almost no exceptions. Penguins, seals, whales, and fish are all examples.

Penguins are undoubtedly the top of the pops Antarctic animal, figuring in children's books, Hollywood films, chocolate biscuits, and the name and logo of a publishing house, among others. Often depicted as funnily clumsy, they are an easy sell in nature documentaries, while news of climate change disrupting the poles usually portrays them as involuntary martyrs facing an uncertain future.[12] This charismatic portrayal of penguins, however, is a relatively recent development.

In *The Worst Journey in the World,* Apsley Cherry-Garrard recalls Robert Falcon Scott's last expedition to the South Pole and its tragic end, in 1911. The youngest member of the team, the author offers vivid impressions of the Antarctic environment and its non-human inhabitants. Among these impressions, his portrayal of Adélie penguins is memorable:

> *The life of an Adélie penguin is one of the most unchristian and successful in the world. The penguin which went in for being a true believer would never*

[12] See U.S. Fish and Wildlife Service, 'U.S. Fish and Wildlife Service Provides Endangered Species Act Protections for Emperor Penguin', 25 October 2022, https://www.fws.gov/press-release/2022-10/emperor-penguin-gets-endangered-species-act-protections; Aaron Sidder, 'Antarctica Could Lose Most of Its Penguins to Climate Change', *National Geographic,* 29 June 2016, https://www.nationalgeographic.com/animals/article/adelie-penguins-antarctica-climate-change-population-decline-refugia. Note, however, that not all penguins are equal. It is a well-known fact that, with climate change, specialists like Emperor and chinstrap penguins will be deeply disrupted, while generalist Gentoos will find a way to adapt and might even flourish. See Kelton W. McMahon, Chantel I. Michelson, Tom Hart, and Michael J. Polito, 'Divergent Trophic Responses of Sympatric Penguin Species to Historic Anthropogenic Exploitation and Recent Climate Change', *Proceedings of the National Academy of Sciences* 116, no. 51 (2019): 25721–27, https://doi.org/10.1073/pnas.1913093116.

stand the ghost of a chance. Watch them go to bathe. Some fifty or sixty agitated birds are gathered upon the ice-foot, peering over the edge, telling one another how nice it will be, and what a good dinner they are going to have. But this is all swank: they are really worried by a horrid suspicion that a sea-leopard is waiting to eat the first to dive. The really noble bird, according to our theories, would say, 'I will go first and if I am killed I shall at any rate have died unselfishly, sacrificing my life for my companions'; and in time all the most noble birds would be dead. What they really do is to try and persuade a companion of weaker mind to plunge: failing this, they quickly pass a conscription act and push him over. And then—bang, helter-skelter, in go all the rest.[13]

This depiction of penguins as amoral and opportunist creatures probably facilitated treating them as objects of scientific wonder at best and food for men and dogs at worst during the 'heroic era' of Antarctic exploration. Since James Cook's tournée around the world in 1774–75, penguins had become the staple food for those venturing to Antarctica, together with other seabirds, like cormorants and albatrosses. As reported by Cook himself, 'the people tired of eating penguins and young shaggs'.[14]

A worse fate (at least in terms of numbers) awaited other Antarctic marine animals, like seals and whales. From the mid-eighteenth century until the mid-twentieth century, these animals fuelled nascent global capitalism with their oil, blubber, and furs.[15] As mentioned before, those killing the animals took a short-term, profit-focused perspective and kept wiping out entire colonies as they advanced further south, acting as if there were limitless abundance.[16]

After British sealers arrived in South Georgia in 1786, others quickly followed. By 1810, the islands were one of the most important centres for sealing

[13] Apsley Cherry-Garrard, *The Worst Journey in the World* (Basic Books, 2008), 581.

[14] Quoted in Maddison, *Class and Colonialism*, 23. When elephant seal populations diminished after years of ruthless exploitation, penguins were sought for their oil in sub-Antarctic waters. The utmost entrepreneur was Joseph Hatch, who installed The Macquarie Island Oil Industry on Macquarie Island, south of Tasmania. In three decades (from the end of the 1880s to 1919), three million king penguins and royal penguins were thrown dead or alive into 'digestors', each of them rendering half a litre of blubber to be used for twine and rope manufacture. The digestors were coal-fired boilers installed amid the birds' colonies. Geoff Chapple, 'Harvest of Souls', *New Zealand Geographic*, July–August 2005, https://www.nzgeo.com/stories/harvest-of-souls/. Public outrage and faltering profits put an end to a business that never reached Antarctica but reflects how these animals were conceived of: primarily as carriers of monetary value to be extracted in whichever way, never mind their well-being.

[15] Elephant seal oil was used for lighting, machine lubrication, and paint manufacture, among other things.

[16] Howkins, *Polar Regions*, 58ff; Maddison, *Class and Colonialism*; Cruz et al., 'The Case of Sealing'.

in the Southern Ocean. Between 1815 and 1817, one British vessel took 5,000 fur seal pelts and 3,500 barrels of elephant seal oil.[17] During the 1821–22 season, there were over forty sealing ships around the South Shetlands and the sealer and explorer James Weddell calculated that, during only two seasons, 320,000 fur seals and 100,000 pups had been killed.[18] It did not help that the sealers worked for a share of the profits, which incentivized even more extraction. During the height of exploitation, 'cooperation within the labour process made it possible to convert millions of seals into bales of skins and barrels of oil in a relatively short period of time'.[19] Given their overexploitation, subsequent cycles (from 1830 to 1850 and from 1870 to 1890) were not as abundant.[20]

As for whales, between 1900 and 1950 it is estimated that nearly a million fin whales, sperm whales, and blue whales were killed in Southern waters. Hunted at a time when 'sustainable' as a concept to designate environmentally responsible behaviour was not in the dictionary, the different species were killed in cycles: 'As one species began to dwindle in abundance, another would be targeted to take its place, and typically the species was the next size smaller'.[21]

Not surprisingly, the twentieth century saw the near extinction of some of these species, as well as the first efforts towards their conservation. Starting with the International Convention for the Regulation of Whaling (hereafter Whaling Convention) in 1946 and following with the signature of the Antarctic Treaty in 1959 as well as some specific conventions and regulations, the once most-exploited Antarctic fauna became the object of stringent protection measures. Under the Convention for the Conservation of Antarctic Seals (1972), six species became fully protected south of 60 degrees south.[22] More recently, in 2001, the Agreement on the Conservation of Albatrosses and Petrels was established to ensure that the thirty-one species present on the continent were as little affected as possible by Southern Ocean fisheries and plastic pollution.[23] What is remarkable is that these agreements refer to the animals to be protected exclusively as resources for human use and benefit: their rationale

[17] Howkins, *Polar Regions*, 61.

[18] Ibid.

[19] Maddison, *Class and Colonialism*, 109. Later, seals also became the standard dinner for explorers and their dogs. Scott calculated that one seal could feed 37 people for two days: ibid., 178.

[20] Ibid.

[21] Robert C. Rocha, Phillip J. Clapham, and Yulia V. Ivashchenko, 'Emptying the Oceans: A Summary of Industrial Whaling Catches in the 20th Century', *Marine Fisheries Review* 76, no. 4 (2014): 37–48, 45.

[22] These were the Southern elephant seal (*Mirounga leonine*), leopard seal (*Hydrurga leptonyx*), Weddell seal (*Leptonychotes weddelli*), Crabeater seal (*Lobodon carcinophagus*), Ross seal (*Ommatophoca rossi*), and Southern fur seals (of the genus *Arctocephalus*).

[23] 'Agreement on the Conservation of Albatrosses and Petrels (2001)', amended 7–11 May 2018, https://www.acap.aq/agreement-text/206-agreement-on-the-conservation-of-albatrosses-and-petrels/file.

is ultimately about protecting the business, where protecting the animals is a means to that end.

The Whaling Convention is explicit on this point, stating that its main goal 'is to achieve the optimum level of whale *stocks* as rapidly as possible without causing widespread [human] economic and nutritional distress'.[24] The prospect of stopping whaling altogether was not on the table in 1946, when the convention stated that whaling operations would continue, focused on 'those species best able to *sustain exploitation* in order to give an interval for recovery to certain species of whales now depleted in numbers [before allowing for their exploitation again]'.[25] The Seal Convention also uses the language of 'stocks' and refers to seals as 'an important *living resource* in the marine environment which requires an international agreement for its effective conservation'.[26] The 'resource' should 'not be depleted by overexploitation' and not 'exceed the levels of the *optimum sustainable yield*'. Exploitation should be placed 'on a rational basis', and science should be mobilized to get the information needed to decide what that rational basis could sustain.[27] Furthermore, although the Albatross Agreement recognizes the importance of albatrosses for some Indigenous cultures (acknowledging that these animals have non-instrumental value for some human groups), the main reason to protect these birds is that they are an 'integral part of marine ecosystems which must be conserved for the benefit of present and future generations'.[28] Albatrosses are thus conceptualized as necessary cogs in an ecological wheel that needs to keep turning for the good of present and future humans.

The final and best example of this instrumental approach to Antarctic animals is the Convention for the Conservation of Antarctic Marine Living Resources or CCAMLR (hereafter, the Marine Convention). Although it states that its main objective is conservation, here conservation 'includes rational use'.[29] Since its inception in 1982, there has been a discussion around how to interpret conservation, rational use, and their relationship, but the very fact that animals are referred to as 'resources' speaks for itself. Even in the most environmentally friendly reading of these terms, the motivation to care for

[24] International Whaling Commission, 'International Convention for the Regulation of Whaling', 2 December 1946, 1, https://archive.iwc.int/pages/view.php?ref=3607&k=, my emphasis.

[25] Ibid., my emphasis.

[26] Antarctic Treaty System, 'Convention for the Conservation of Antarctic Seals (1972)', 137, https://documents.ats.aq/keydocs/vol_1/vol1_13_CCAS_CCAS_e.pdf, my emphasis.

[27] Ibid.

[28] 'Agreement on the Conservation of Albatrosses and Petrels'.

[29] Commission for the Conservation of Antarctic Marine Living Resources, 'Convention for the Conservation of Antarctic Marine Living Resources', Article II, 1982, page last modified 25 November 2013, https://www.ccamlr.org/en/organisation/camlr-convention-text.

Antarctic marine animals is ultimately to maximize their use for us. It is thus not surprising that many have criticized the Marine Convention as a fishing convention.[30]

This instrumental approach is so ingrained that it can be found even in the places where one would least expect it. In the jaw-dropping BBC Earth documentary *Antarctica 3D*, Antarctic marine animals are presented as worth protecting because of their role in carbon sequestration in the Southern Ocean and, therefore, in mitigating global climate change. From being indispensable to the industries that partly created anthropogenic climate change, they have now become indispensable to protect us from the effects of those very same industries.

5.2 In the name of the penguins

One of the least faunistically diverse places on earth ('exceptional in its simplicity'[31]), Antarctica has no terrestrial mammals, reptiles, or amphibians. However, partly thanks to the conventions mentioned above, the numbers of the most sought and exploited marine animals of the past are slowly recovering. There are six species of penguin in Antarctica, five of whom breed there: Emperor, Adélie, Macaroni, Chinstrap, and Gentoo. Their estimated population is 12 million (around 5.77 million breeding pairs), the largest group being the Adélie penguins (with over 4 million individuals) and the smallest being the Macaroni (around 13,000). Among the six species of seals (Ross, Weddell, Crabeater, leopard, fur, and elephant), the Crabeaters are the most numerous, with approximately four million individuals. Together, they add up to around 5.5 million individuals. Two hundred and fifty thousand whales from seven different species (killer, sperm, humpback, Southern right, Antarctic blue, Sei, Antarctic minke, and fin) are also estimated to navigate through Antarctic waters during their yearly migration.

[30] For a defence of the interpretation that the Marine Convention is not about fishing, see Anthony J. Press, Indi Hodgson-Johnston, and Andrew J. Constable, 'The Principles of the Convention on the Conservation of Antarctic Marine Living Resources: Why Its Commission Is Not a Regional Fisheries Management Organization', in *Governing Marine Living Resources in the Polar Regions*, ed. Nengye Liu, Cassandra M. Brooks, and Tianbao Qin (Edward Elgar Publishing, 2019) and Australia and USA, 'Conservation at CCAMLR: Understanding Article II of the Convention on the Conservation of Antarctic Marine Living Resources', CCAMLR-XXXV/BG/28, 2016, https://www.antarctica.gov.au/site/assets/files/47746/camlr-art-ii-interpretation-final.pdf.

[31] Peter Convey and Sandra J. McInnes, 'Exceptional Tardigrade-Dominated Ecosystems in Ellsworth Land, Antarctica', *Ecology* 86, no. 2 (2005): 519–27, https://doi.org/10.1890/04-0684.

When it comes to seabirds, over 100 million albatrosses, petrels, and skuas (among 20 other species of bird) breed around the Antarctic coastline and the offshore islands.[32] One of the two largest colonies of Antarctic Petrels, in Svarthamaren (Queen Maud's Land), hosts 100,000 breeding pairs.[33]

As for fish, seven species are considered endemic to the Southern Ocean and around 200 have been registered south of the Antarctic Convergence, half of them demersal fish, living at the bottom of the ocean.[34]

But the most populous inhabitant of Antarctic waters is without doubt krill. There are five different species, the dominant of which is Antarctic krill (*Euphasia superba*), with an estimated biomass of 400 million tonnes, roughly equivalent to the weight of all humans on earth.[35]

In interpreting where the intrinsic value of Antarctica lies, one plausible answer is to locate it in the millions of sentient beings that inhabit the continent. If this is the case, shouldn't the normatively relevant interests of these animals be considered when making decisions about Antarctica? If so, how ought those interests to be represented?

Alfonso Donoso answers the first question in the affirmative and provides a sketch of an answer to the second one. All sentient animals in Antarctica should be considered as holders of the intrinsic value that the Environmental Protocol commits itself to protecting. In line with other animal ethicists, Donoso takes sentience to be the necessary and sufficient condition for being morally (and, by extension, legally and politically) considered.[36] To the extent that their lives can go better or worse, the interests of sentient animals are relevant. At the minimum, human agents ought not to harm them, and ought not to allow practices and policies conducive to harming them.[37]

Donoso claims that this is not what happens in Antarctica today, where the basic interests of non-humans continue to be trumped by non-basic human

[32] Australian Antarctic Program, 'Flying Birds', last updated 12 August 2010, https://www.antarctica.gov.au/about-antarctica/animals/flying-birds/.

[33] Arnaud Tarroux, Per Fauchald, and Øystein Varpe, 'An Ocean of Possibilities: Antarctic Petrels' Foraging in the Southern Ocean', The Fram Centre, published 18 January 2021, https://framsenteret.no/forum/fram-forum-2021/an-ocean-of-possibilities-antarctic-petrels-foraging-in-the-southern-ocean/.

[34] Norsk Polarinstitutt, 'The Fauna of Antarctica', https://npolar.no/en/themes/fauna-antarctica/.

[35] Australian Antarctic Program, 'Antarctic Krill', last updated 28 February 2025, https://www.antarctica.gov.au/about-antarctica/animals/krill/. Consistent with their treatment as resources, krill get measured by their collective weight rather than by individuals. If it comes to individuals, the number is astronomical: 800 trillion, that is, 100,000 for each human being on earth.

[36] The classic reference to sentience as a morally relevant characteristic is Singer, *Animal Liberation*.

[37] Alfonso Donoso, 'Valor intrínseco en la Antártica', *Revista de Estudios Hemisféricos y Polares* 12, no. 1 (2021): 44–57, 52. See also Alfonso Donoso, 'Intrinsic Value, Sovereignty Rights, and Sentient Animals in Antarctica', *The Polar Journal* 14, no. 2 (2024): 373–88, https://doi.org/10.1080/2154896X.2024.2415809.

interests. As he points out, it is not clear how the intrinsic value of individual seals, penguins, and other marine animals can be reconciled with some parts of the Environmental Protocol and, especially, with the content of some of its six annexes. Take, for example, Annex II, on the Conservation of Antarctic Fauna and Flora, which states that 'Taking or harmful interference shall be prohibited, except in accordance with a permit'. This permit can be issued for a variety of reasons, among them some quite trivial ones:

> *(a) to provide specimens for scientific study or scientific information; (b) to provide specimens for museums, herbaria, zoological and botanical gardens, or other educational or cultural institutions or uses; and (c) to provide for unavoidable consequences of scientific activities not otherwise authorized . . . or of the construction and operation of scientific support facilities.*[38]

If we took the interests of Antarctic sentient animals seriously, we would not see them as potential 'specimens', let alone display their bodies, dead or alive, thousands of miles away from their homes just for the purposes of human *Bildung*. Neither would we get to discuss so lightly how many penguin and seal lives an airstrip is worth, even if built for the interests of science.[39]

Donoso is not primarily interested in the constitution of the Antarctic demos, but rather in the question of what 'Antarctica' means in Article 3 of the Environmental Protocol, and how it should be implemented as a result. However, one could see these inquiries as overlapping. Article 3 of the Environmental Protocol, on the one hand, calls for protecting the intrinsic value of Antarctica, which is—following Donoso—the value possessed by all Antarctic sentient beings. On the other hand, if one grants that the best candidate for delimiting the demos in Antarctica and elsewhere is the *All Affected Interests Principle*, as discussed in Chapter 4, then all Antarctic sentient beings, so long as they may be relevantly affected by human actions and/or omissions, should be considered in the process of decision-making.[40] At least

[38] Antarctic Treaty System, 'Annex II to the Protocol on Environmental Protection to the Antarctic Treaty: Conservation of Antarctic Fauna and Flora (1991)', Article 3, paras 1 & 2, https://documents.ats.aq/recatt/Att432_e.pdf.

[39] See, for example, the controversial Australian project to build an airstrip that would have affected thousands of these animals: Shaun Brooks and Julia Jabour, 'Australia Wants to Build a Huge Concrete Runway in Antarctica. Here's Why That's a Bad Idea', *The Conversation*, 17 July 2020, https://theconversation.com/australia-wants-to-build-a-huge-concrete-runway-in-antarctica-heres-why-thats-a-bad-idea-139596.

[40] The argument works equally well if the principle chosen is the *All Subjected Interests Principle*: after all, we are discussing whether those living in the territory and being subjected to its laws should be considered in the decision-making process.

when it comes to the Environmental Protocol, it is clear that the rules it establishes affect those sentient beings directly and in relevant ways. Therefore, the Environmental Protocol's signatories—especially those who hold territorial claims, Donoso suggests—ought to show a serious commitment to the protection of those sentient beings in order to legitimize their own claims.[41]

Donoso unnecessarily ties the legitimacy of the frozen territorial claims to the protection of Antarctic animals. As I have pointed out earlier, the territorial claims should be given up once and for all, and no action performed on the continent should give individual states grounds for making new territorial claims. Instead, a more fruitful strategy would be to argue that any signatory state, claimant or not, should have the legitimacy of their engagement measured by how well they manage to fulfil the goals of the Protocol. Going back to the example above: for states to really show that they have committed themselves to the fulfilment or at least the non-violation of its environmental principles, they should protect all sentient beings in Antarctica in a way such that non-basic human interests do not trump basic non-human interests. This would require them to work towards an amendment of (at least) the content of some appendixes and—more ambitiously—towards a major reinterpretation of the Environmental Protocol and of how its goals ought to be pursued, as I argue in Chapter 8. It would also require reinterpreting the text of the Marine Convention, such that the definition of 'rational use' becomes much more demanding than it is today, when the Convention sometimes dangerously resembles a fisheries management instrument.[42] Moreover, if fish and krill are sentient, and if intrinsic value is located in sentient beings, then the whole Antarctic fishing industry should come under scrutiny. The Southern Ocean's less charismatic, yet sentient, inhabitants should become protected too.[43]

The interests of Antarctic animals could then be encapsulated within the interests of the states or, more precisely, their delegates. Robert E. Goodin has suggested that considering nature's interests requires humans to *internalize*

[41] Donoso, 'Valor intrínseco en la Antártica', 54.

[42] Jennifer Jacquet, Eli Blood-Patterson, Cassandra Brooks, and David Ainley, '"Rational Use" in Antarctic Waters', *Marine Policy* 63 (2016): 28–34, https://doi.org/10.1016/j.marpol.2015.09.031.

[43] There is a big gap on species-specific welfare knowledge, but there is accumulating evidence that fish are sentient creatures. See, for example, Becca Franks, Christopher Ewell, and Jennifer Jacquet, 'Animal Welfare Risks of Global Aquaculture', *Science Advances* 7, no. 14 (2023): eabg0677, https://doi.org/10.1126/sciadv.abg0677. Regarding krill and similar invertebrates, the issue of sentience is understudied to say the least. See Jonathan Birch, 'Animal Sentience and the Precautionary Principle', *Animal Sentience* 2, no. 16 (2017), https://doi.org/10.51291/2377-7478.1200. An alternative argumentative path that I do not explore here is to say that what matters is not whether marine animals are sentient or not, but whether they *resist*: Dinesh Joseph Wadiwel, 'Do Fish Resist?', *Cultural Studies Review* 22, no. 1 (2016): 196–242, https://doi.org/10.5130/csr.v22i1.4363.

those interests and reflect them in political deliberation and decision-making.[44] This form of 'encapsulated representation' does not have a very reputable history when done on behalf of other rational and autonomous adult humans, like masters representing their slaves and husbands representing their wives. However, according to Goodin, it is the best way to go when those whose interests should be protected do not have the agency required to participate directly in the political process. Children are a good example: even though they don't have the right to vote, we assume that most of the time their parents will be responsive to their interests and will be their best spokespersons until they can speak for themselves. Here, similarly, the idea would be to let nature—in this case, sentient animals—be represented by humans:

> *There is simply no other way in which nature's interests can find political representation except through being politically incorporated within the interests of sympathetic humans capable of bringing political pressure to bear on nature's behalf.*[45]

Along these lines, the consultative parties should ensure that their Antarctic representatives show sympathy towards Antarctic animals in a way that is reflected in the decisions, measures, and resolutions that constitute the Treaty's legal corpus. This would probably require a different way of selecting, educating, and briefing those who end up participating in the Antarctic Treaty Consultative Meetings. Ideally, they would have to be enlightened bureaucrats with a special sensitivity for and knowledge of sentient Antarctic fauna. Inspired by Adam Smith's figure of the 'impartial spectator', they should act in a way that is non-biased, well-informed, and attentive towards all the relevantly affected parties.[46]

One could object, however, that having the usual suspects speak for sentient Antarctic animals will not work. After all, delegates to the meetings are there mainly to represent the interests of their fellow human citizens at best, and of the government of the day at worst. These interests might well be at odds with those of the animals—especially if the humans or governments represented still see these animals as resources to be fished, hunted, skinned, and savoured. Because humans have a voice, whereas the animals lack it, in practice this will

[44] Goodin, 'Enfranchising the Earth'.
[45] Ibid., 844.
[46] Adam Smith, 'The Theory of Moral Sentiments', in *The Glasgow Edition of the Works and Correspondence of Adam Smith*, ed. A. L. Macfie and D. D. Raphael, vol. 1 (Liberty Fund, 1982), III.I.2, 110.

mean that the interests of the latter will not weigh as heavily as the interests of the former—if they are weighed at all. Moreover, as Alasdair Cochrane points out, the historical exclusion of animals' interests from the political debate makes their proper representation ultimately contingent on the goodwill of the decision-makers.[47] What are needed, instead, are independent representatives.

5.3 Representing wild animals

Instead of keeping only states (through their representatives) as deliberators and decision-makers, one could propose that Antarctic animals should have their own spokespersons participating in the negotiations that ultimately determine their present and future. This could be inspired by what Sue Donaldson and Will Kymlicka have called a 'form of proxy representation by human beings who are committed to the principle of animal sovereignty'.[48]

Since they published *Zoopolis* in 2011, Donaldson and Kymlicka initiated a 'political turn' in animal ethics.[49] Instead of thinking of our relations with animals as lying beyond the political realm, what they suggest is that different kinds of animals ought to have different political statuses. In a nutshell, while domesticated animals should be considered citizens, wild animals (like those found in Antarctica) should be treated as foreign peoples in their sovereign territories, and our dealing with them should be regulated by the same principles that govern our dealings with foreign nations.[50]

For Donaldson and Kymlicka, wild animal communities have an interest in maintaining their own models of social cooperation and organization, and in leading their lives without our interference. This grounds a right to self-determination, just like it does in the case of human communities. Although they don't elaborate on how to implement their proposal (they mention implementation in the closing paragraphs of the relevant chapter, only to say that it deserves further research), a representative of Antarctic animals faithful to *Zoopolis* would be roughly guided by the following principles: First, if wild

[47] Alasdair Cochrane, *Sentientist Politics: A Theory of Global Inter-Species Justice* (Oxford University Press, 2018), 43.

[48] Sue Donaldson and Will Kymlicka, *Zoopolis: A Political Theory of Animal Rights* (Oxford University Press, 2011), 209.

[49] Tony Milligan, 'The Political Turn in Animal Rights', *Politics and Animals* 1 (2015): 6–15; Alasdair Cochrane, Robert Garner, and Siobhan O'Sullivan, 'Animal Ethics and the Political', *Critical Review of International Social and Political Philosophy* 21, no. 2 (2018): 261–77, https://www.tandfonline.com/action/showCitFormats?doi = 10.1080/13698230.2016.1194583.

[50] There is a third category, liminal animals, who live around us and whom we merely tolerate, without making them part of our political community.

animals are sovereign over their territory, humans should stop invading, colonizing, and robbing it. Halting further destruction and exploitation of their habitat should be the priority. Second, to cooperate on a basis of equality and non-exploitation, human activities should be constrained in areas of overlapping sovereignty—in this case, mainly, the seas and coastal areas. Depending on how far down the evolutionary tree one is willing to grant sentience, this would have profound effects on fishing regulations and, therefore, on the Antarctic Marine Convention. Third, leaving them to lead their lives without human interference does not mean that humans would have no positive duties towards sentient animal communities in Antarctica. In cases of natural or human-made disasters or external threats, where the very existence of the community is jeopardized, humans should assist them to the best of our abilities to let them 'get back on [their] own feet' (or flippers).[51] Donaldson and Kymlicka's examples of threats are 'a rogue bacterium, or giant meteor'.[52] Two examples in Antarctica could be the dramatic effects that climate change is having on the breeding and food habits of some species, and the arrival of highly pathogenic avian influenza with its devastating mortality rates.[53]

To sum up the discussion so far, if we grant that Antarctic sentient beings have relevant interests and take the *All Affected Interests Principle* as the one that should be used to bound the demos, it follows that Antarctic sentient beings should be considered as part of the Antarctic political community. One way to realize this, along the lines of Donoso and Goodin, would be to ensure that the delegates from states that participate in Antarctic deliberation internalize those interests and commit themselves to their protection. Contrary to what happens today, this would require that the basic interests of those animals should not be subordinated to non-basic human interests—like wanderlust, gustatory explorations, and, arguably, even scientific inquisitiveness. Another way to realize this, inspired by Donaldson and Kymlicka, would be to have proxy representatives for Antarctic wild animal communities, who ensure that the minimal principles that guide relationships between sovereign entities are respected: non-interference (unless there is a case to be made for emergency assistance), equality, and non-exploitation. These representatives could sit

[51] Donaldson and Kymlicka add here the condition that 'our efforts to assist are not rejected': Donaldson and Kymlicka, *Zoopolis*, 181. How that is to be judged is not spelled out.

[52] Ibid., 206.

[53] Sofia Moutinho, 'H5N1 Avian Flu Is Spreading Rapidly in Antarctica', *Science* (United States) 387, no. 6739 (2025): 1130–31, https://doi.org/10.1126/science.adx3891; and British Antarctic Survey, 'Highly Pathogenic Avian Influenza (HPAI) Has Been Confirmed in Brown Skua Populations on Bird Island, South Georgia—The First Known Cases in the Antarctic Region', 23 October 2023, https://www.bas.ac.uk/media-post/first-confirmed-cases-of-avian-influenza-in-the-antarctic-region/.

among other states' delegates in the Antarctic Treaty Meetings and other fora that concern them.

Both the encapsulated representation model and the proxy representation model, however, downplay a key challenge; namely, they assume that it is a relatively straightforward matter to interpret the relevant interests of the animals and to find the best means for their protection (or, at least, the best means to avoid what would be the most obviously damaging measures).[54] Whereas in the case of humans, it is individuals themselves (on their own or as a part of a collective) who can clarify their interests and demand what should be done accordingly, here we can only trust our own judgements in the matter—however impartial, well-informed, and attentive these judgments try to be. This is a key challenge, because if we prioritize the basic interest of individual animals in not suffering and/or not getting killed, we will arrive at distinct and even contradictory prescriptions from those that would result from prioritizing the interests of the animals in being part of an autonomous community (assuming that it makes sense to talk of an autonomous animal community). Furthermore, regarding how we end up conceptualizing the relevant interests, one could think that the representatives could be assigned by species, by ecological or geographical region, or by distinguishing between permanent and migratory residents—among other options.[55]

Against this challenge, in the next section I suggest that, on the one hand, the idea of a wild animal community leads in practice to very similar prescriptions to those suggested by advocates of more ecocentric approaches (which I discuss in more detail in Chapter 6). On the other hand, if the aim is to protect the individual interests of wild animals in not suffering and/or getting killed, this can quickly lead to conclusions that would demand a huge shift from our current moral, political, and legal paradigms—to the point of being accused of absurdity.

At this point, some might wonder why I am highlighting these two approaches to animals' interests while ignoring others that are also plausible to hold. For example, one could adopt Clare Palmer's laissez-faire position towards wild animals, which is more hands-off than Donaldson and Kymlicka's and is not based on the value of upholding collective self-determination.[56] Or one could endorse Alasdair Cochrane's position, where wild animals are

[54] Uncertainty about their interests is, in fact, one of the factors mentioned to explain why animals tend to be politically disregarded: Pablo Magaña, 'The Political Representation of Nonhuman Animals', *Social Theory and Practice* 48, no. 4 (2022): 665–90, 688, https://www.jstor.org/stable/48747320.

[55] I thank Alfonso Donoso for suggesting this point.

[56] Clare Palmer, *Animal Ethics in Context* (Columbia University Press, 2010), chapter 4.

recognized as equal members in our political communities, towards whom we have some duties of assistance, but no duties of massive intervention.[57] The reason why I focus on these two views—which one might call the sentientist and the communitarian—is that they sit at opposite ends of the spectrum, they are both well supported by arguments, they are both held by specialists in animal ethics and animal political theory, and they lead to contradictory implications. This, in my view, shows that deciding on a representation model for non-humans is anything but straightforward. The same exercise could be done starting from other views located between these two, and the outcome would be similar.

5.4 A community red in tooth and claw

The standard story to justify the right of political self-determination for human collectives is that part of what it means to be an autonomous human individual is to be able to participate in collective decision-making processes that ultimately shape our lives. Individual self-determination is thus boosted by collective self-determination. Moreover, it is thought that giving human collectives rights to determine their own rules is purportedly the best way to ensure that the most important interests of the individuals living in that community will be protected and fulfilled. The protection and fulfilment of basic interests and the furthering of individual autonomy are thus seen as justifications for the rights of group self-determination. Moreover, control over the territory where the group lives is thought to be key among these rights: those who inhabit the place must be able to decide how to rule it, use it, distribute it, and so on, for the achievement of the above aims. So long as they do not harm others, they should not be interfered with in this process.[58]

At first sight, it is tempting to draw an analogy and think about animals in the wild as sovereign, self-determining collectives too. As Donaldson and Kymlicka put it, 'wild animals should be seen as members of their own animal communities living on their own territory, and hence as having sovereignty rights over themselves and their territory'.[59] As such, the authors continue, they have claims of justice against humans not to be harmed or killed for our benefit. If the purpose of sovereignty, as they see it, is to protect basic interests

[57] Cochrane, *Sentientist Politics*, chapter 5.
[58] See, for example, Moore, *Political Theory of Territory* and Stilz, *Territorial Sovereignty*.
[59] Donaldson and Kymlicka, *Zoopolis*, 144.

against certain standard threats, these interests in the case of wild animal communities are 'in maintaining valued forms of social organization tied to a particular territory against the threat of conquest, colonization, displacement and alien rule [by humans]'.[60]

The attractiveness of this position quickly evaporates, however, once we ask questions like: Who exactly are the members of these animal communities? Are they all from the same species, or from species that coexist harmoniously together? Do predator and prey inhabit the same, or different, animal communities? Why do wild animal communities have claims of justice only against human-inflicted harm, and not also against non-human inflicted harm, if the latter may be prevented at a reasonable cost? Why are the standard threats against which sovereignty should protect individuals limited to external (human) threats in the case of wild animal communities? And what about internal threats, arising from other members of the same community? Ultimately, what are the basic interests of wild animals and is the best way to protect these interests to grant them sovereign rights?

These questions have been posed to Donaldson and Kymlicka in different forms before, and they have responded to them.[61] But their responses require us either to adopt a controversial notion of relevant animal interests or to move closer to an ecocentric position.

To the question of whether one can meaningfully talk about 'wild animal communities', the authors respond that, while it would be strange (or, rather, ecocentric) to say that predator and prey belong to the same community, it is plausible to understand the bearer of sovereignty to be 'multi-species habitats or eco-regions'.[62] Insofar as predator and prey are both dependent on the same habitat, they continue, 'animals within the boundaries of these territorial regions have a common interest in protecting this habitat from external threats'.[63]

Here, however, Donaldson and Kymlicka are equivocating on the meaning of 'common interest'. When we say that a collection of individuals has a *common interest*, the underlying thought is that they are pointing together towards a common goal. Human societies, for example, have a common interest in living peacefully and safely, and therefore set rules to ensure that these ends are met for all their members as much as possible. Donaldson and Kymlicka, instead,

[60] Ibid., 151.

[61] See, for example, Alasdair Cochrane, 'Cosmozoopolis: The Case Against Group-Differentiated Animal Rights', *Law, Ethics and Philosophy* 1 (2013), 127–41; Oscar Horta, 'Zoopolis, Interventions and the State of Nature', *Law, Ethics and Philosophy* 1 (2013): 113–25.

[62] Sue Donaldson and Will Kymlicka, 'A Defense of Animal Citizens and Sovereigns', *Law, Ethics and Philosophy* no. 1 (2013): 143–60, 152.

[63] Ibid., 152.

seem to confuse having *a common interest* with having *the same interest*: soldiers in opposite trenches have the same interest (namely, not getting killed by those on the other side), but that does not turn them all into one group united by a common interest. Analogously in the case of multispecies habitats or eco-regions, Adélie penguins and leopard seals might have the same interest in the territory, but this does not mean that they have a common interest in being in the place together. Talk of 'wild animal communities', from this perspective, is no more than a bad and even damaging metaphor. No Adélie penguin in his right mind would—if asked—suggest that the leopard seal be his fellow citizen. What kind of community would this be, where the strongest kill the weakest in a constant struggle for survival?

Against this kind of objection, Donaldson and Kymlicka point out that one should understand 'community' in a minimally demanding sense and recognize that there are mutual dependencies among animals 'even though individual relationships might be violent and competitive'.[64] The only standard threats that they think should be ruled out as a matter of justice are therefore external, human threats to the community rather than internal threats to individual members. For this reason, they are against intervening in predator–prey relationships on the grounds that this would radically disrupt their ways of life, restricting their freedom and autonomy, 'including the freedom to fully develop and explore one's capabilities [also known as flourishing]'.[65] It would turn nature into a zoo where humans would manage their movements, their reproduction, and their socialization.[66]

But this response relies on an attitude towards wild animal interests where the most basic interest in not starving or getting eaten alive gets downplayed and ultimately ignored. The fact that these animals cannot choose their diet and cannot fight their instincts (thus making them devoid of moral agency) does not change the elementary truth. If we cared for the well-being of non-human animals in exactly the same way that we care for the well-being of humans, we would think it problematic, to say the least, to let human individuals lacking moral agency interact with each other undisturbed by us. Donaldson's and Kymlicka's claim that the way of life of wild animals should be kept as it is may be accused of being a romanticization of reality in the wild.

Another way to inquire into the basic interests of wild animals is to go back to the very same quality that many animal theorists—including Donaldson

[64] Ibid., 154.
[65] Donaldson and Kymlicka, *Zoopolis*, 166.
[66] Donaldson and Kymlicka, 'Defense', 155.

and Kymlicka—consider as the necessary and sufficient condition for being a subject of moral consideration; as mentioned before, sentience. If we look at history, one of the main external challenges that humans have faced is predation. *Little Red Riding Hood* may be a children's story, but it reflects a very deep-seated human worry: sentient beings don't like getting eaten alive. As Val Plumwood reflected on her near-death experience of being dragged down three times by a saltwater crocodile in Kakadu, Australia:

> *This was a strong sense, at the moment of being grabbed by those powerful jaws, that there was something profoundly and incredibly wrong in what was happening, some sort of mistaken identity. My disbelief was not just existential, but ethical . . . The creature was breaking the rules, was totally mistaken, utterly wrong to think I could be reduced to food. As a human being, I was so much more than food.*[67]

We don't regard it as desirable, or even as an inevitable part of our way of life, to get eaten by others—as much as we may find it desirable and inevitable to be predators ourselves (but that is the topic for another discussion). This most basic interest in not being eaten alive is shared with other animals, including wild ones. 'Nobody likes to get eaten' is not a controversial claim. Moreover, seen from the predator's perspective, 'Nobody likes to starve to death' is also uncontroversial. Why insist then that, sentient as they are, it is in the interest of individual wild animals to stay in predator–prey relationships? Inversely, why insist that they have no interest in putting an end to these relationships? (This strikes me as even more inconsistent if we grant domesticated animals that interest and protect it by right.[68])

[67] Val Plumwood, 'Meeting the Predator', in *The Eye of the Crocodile*, ed. Lorraine Shannon (ANU E Press, 2012), 11–12, her emphases. Being through that traumatic experience, for Plumwood, showed her our mistaken perception of our human place in nature: instead of denying our status as potential food for others, we ought to embrace it—not by throwing ourselves willingly into the jaws of the crocodile but by acknowledging our inclusion in the animal and natural world.

[68] Imagine that the following experiment could be performed. Every morning, an Adélie penguin is given two options: she can either jump into the unknown and hope for the best while she feeds on fish (like in Cherry-Garrard's story), or she can jump into predator-free waters, where she can get lab-produced fish pellets. Assuming that the fish pellets are palatable enough, how long would it take the penguin to realize that the second option is preferable than the first? The answer is even clearer, I think, if the animals who are given the choice are vegetarian rather than carnivorous or omnivorous. Imagine that every morning a reindeer is given the option of grazing in lands inhabited by wolves, or in lands with no predators. Would the freedom, autonomy, and potential for flourishing of the reindeer be thwarted if he picked the latter option rather than the former? Would the reindeer mind our managing his circumstances if the cost is a wolfless existence? Questions like these tend to be ridiculed, but they raise genuine moral puzzles that anyone proposing a consistent animal political theory should grapple with.

The obvious answer seems to be that the preservation of certain animal relations in wild territory ultimately matters more than the well-being of the individuals taking part in them. This seems to be confirmed by Donaldson's and Kymlicka's reply to Oscar Horta, who characterizes wild habitats as failed states in need of humanitarian intervention.[69] Far from that, they claim, the only interventions that are acceptable in the wild are micro-level acts of compassion (e.g. if we see an animal suffering and are able to help it without disrupting the habitat), or helping to restore the circumstances that those ways of life are adapted to 'in the face of challenges that overwhelm wild animals'.[70] Among these challenges, interestingly, they mention 'a plague that threatens to wipe out *an entire ecosystem* or efforts to redress human-caused *environmental degradation*'.[71]

The fact that these interventions are geared at keeping the ecosystem rolling suggests that Donaldson and Kymlicka are, *pace* them, treading close to ecocentric territory. Although they are adamant that ecocentrism is not their position, their subordination of basic animal interests to the preservation of the ways of life in the wild, and their acknowledgment that intervention is acceptable if geared at maintaining the circumstances underlying those ways, indicates otherwise. By locating the subject of sovereignty in multispecies habitats or ecoregions, they come very close (if not in principle, at least in practice) to what ecocentric approaches would commend; namely, adopting an attitude as hands-off as possible where we mostly let nature run its course.

Applied to Antarctica, following this line of thought would require much stricter restrictions on scientific and tourist activities and, arguably, it would require putting an end to fishing, the most lucrative economic activity in the Southern Ocean. As I show in Chapter 6, these implications are not far from what a rights of nature approach to Antarctic representation would prescribe. They also seem more aligned with what the best scientific evidence available would recommend, and also with an attitude of epistemic humility that might be the most advisable under the current circumstances. But what would happen if we opted for the sentientist approach instead and agreed that wild territories—including Antarctica—are a terrible idea for animal well-being?

[69] Oscar Horta, 'Debunking the Idyllic View of Natural Processes: Population Dynamics and Suffering in the Wild', *Télos* 17, no. 1 (2010): 73–88.

[70] Donaldson and Kymlicka, 'Defense', 155.

[71] Ibid., my emphases.

5.5 The largest zoo in the world?

Cherry-Garrard's memories of Antarctica are not precisely objective. It is for a reason that he entitled his book *The Worst Journey in the World*. In that journey, he experienced how inhospitable the continent could be for humans, taking the lives of his companions and almost his own. He also observed how harsh Antarctica was towards its own native inhabitants:

> *The Adélie penguin has a hard life: the Emperor penguin a horrible one. Why not kill off the unfit right away, before they have had time to breed, almost before they have had time to eat? Life is a stern business in any case: why pretend that it is anything else? ... But it is grim: Nature is an uncompromising nurse.*[72]

One could charge Cherry-Garrard with anthropomorphizing the penguins and not being able to see that there was a harmony and a logic underlying what he saw as uncompromising nature. Or one could say that his views confirm Horta's argument about wilderness as a moral problem. Why do we assume that the state of nature is something undesirable for humans, as Hobbes pointed out, but desirable (or, at least, not undesirable) for non-humans? Would not true multispecies justice require deep interventions in nature to prevent unnecessary suffering, harm, and death?

In recent years, the idea that predation and life in the wild in general is morally problematic has gained traction among philosophers. Jeff McMahan, for example, thinks that there is a moral reason to help prey from being killed and to prevent other forms of suffering in the wild when this can be done 'at reasonable cost and without neglecting other duties'.[73] McMahan defends this position (often tagged as utilitarian) against four main objections.[74] The first one is that intervening to end predation would be a recipe for ecological havoc. This would require sterilizing or genetically modifying carnivores, so that their descendants become vegan, and actively controlling the numbers of herbivores who, absent any predators, would soon be too many. McMahan agrees that this objection is weighty enough to prevent us from taking measures against

[72] Cherry-Garrard, *Worst Journey*, 584.

[73] Jeff McMahan, 'The Moral Problem of Predation', in *Philosophy Comes to Dinner: Arguments About the Ethics of Eating*, ed. Andrew Chignell, Terence Cuneo, and Matt Halteman (Routledge: 2015), 291.

[74] I say that this position is 'tagged' as utilitarian because arguments can be given to support it that are not necessarily utilitarian. If we think in terms of rights, for example, this position could follow from a duty to rescue, correlated with the animal's right not to suffer and be killed. Or, if we think in terms of capabilities, this position could follow from the idea that wild animals' flourishing is thwarted by predation. See, for example, Martha C. Nussbaum, *Justice for Animals: Our Collective Responsibility* (Simon and Schuster, 2022).

predation at present, but believes that it will become feasible, and therefore morally desirable, in the future. A second argument, about the impersonal value of species that would be lost if we were to get rid of all carnivores, is unpersuasive—he thinks—when we compare it to the value of eliminating suffering for the prey. A third one says that animal suffering does not matter or matters less than human suffering, but this smacks of speciesism (defined as a bias that makes us favour the interests of members of our own species above and against those of other species[75]). Finally, and most relevant for our discussion, is the objection that we should not intervene in wild communities. From the perspective of Donaldson and Kymlicka (to whom McMahan does not refer), the reason for this is that we would undercut animals' freedom. From the perspective of ecologists and environmental philosophers, the reason is that we would be playing God and doing something that is 'biologically preposterous', in J. Baird Callicott's words.[76] As McMahan points out, however, there is no reason why biology should constrain our moral norms in the case of non-human animals, if it doesn't in the case of humans. Moreover, those who hold this kind of position find interventions in the wild acceptable if they restore predator–prey relationships, which shows that they are not against intervention per se: 'Yet the cycle of predation they wish to preserve and prolong indefinitely is, for the prey, a cycle of fear, suffering and violent death'.[77]

Once we see nature this way, the goal of preserving the Antarctic environment quickly loses its lustre. Antarctica becomes less of a sublime landscape and more of an area ravaged by war where, if we don't intervene, it is not because we shouldn't, but because we have good reason to think that the results would be even worse. If we could, then, we should eventually aim at turning Antarctica into the largest zoo on earth. But should we, really?

5.6 Implementing representation

The Environmental Protocol assigns intrinsic value to Antarctica and commits parties to its protection, but does not specify who Antarctica is, why it has intrinsic value, and how that value ought to be protected. From the perspective of animal ethics and animal political theory, one could argue that the sentient inhabitants of Antarctica should be the bearers of that value, and that their basic

[75] Singer, *Animal Liberation*, 6.

[76] J. Baird Callicott, 'Animal Liberation: A Triangular Affair', *Environmental Ethics*, no. 4 (1980): 311–38, 333.

[77] McMahan, 'Moral Problem', 289.

interests should be protected accordingly. So far, the protection of Antarctic animals has been done for purely instrumental reasons. I have therefore suggested two alternative models that would take these animals into account more directly: either having humans with decision-making power internalize or encapsulate animal interests and make them count in the decision-making process, or setting up a proxy model of representation, with independent delegates advocating directly for the animals' interests. However, I have shown that, depending on what interests we prioritize, the resulting prescriptions can be very different and even contradictory. If we care first and foremost for the basic interest of individual Antarctic animals in not suffering and not getting killed, the whole idea of Antarctica as a place worth preserving is put into question: wilderness in general (Antarctica included) is something that we should ideally get rid of. As Oscar Horta affirms, 'we need to question the idea that nature is an idyllic place in which any kind of intervention should be considered taboo'.[78] If, instead, we prioritize the interests of Antarctic animals in roaming undisturbed by humans, we will accept the existence of violent and harmful relationships between them for the sake of preserving their freedom and potential to flourish within these wild communities. The upshot is that, even if we agreed that sentient Antarctic animals ought to be represented, there would not necessarily be unanimity regarding what interests should be the relevant ones requiring protection, and how.

But the challenge of interpreting the relevant interests of those with whom we cannot speak should not deter us from aiming at representing them in those political fora where decisions are made that relevantly affect them. As Danielle Celermajer and colleagues put it when discussing the possibilities of implementing multispecies justice, the inclusion of non-human beings in the political community will always rest 'on a type of second-order exclusion, insofar as beings other than humans must always be represented by, and rely on, the accurate translation of humans'.[79] Or maybe the way to go is to conceptualize representation differently, not as aiming at an accurate translation or mirroring of the interests of the represented, but rather as a statement about the kinds of relations that we wish to cultivate and promote with them. This is Mihnea Tanasescu's argument: we will never find the evidence for what the 'real' interests of the animals are. Instead, as he puts it: 'What knowledge must necessarily leave out is the part that representation fills in, namely the

[78] Horta, 'Debunking the Idyllic View', 87.

[79] Danielle Celermajer, David Schlosberg, Lauren Rickards et al., 'Multispecies Justice: Theories, Challenges, and a Research Agenda for Environmental Politics', *Environmental Politics* 30 (2021): 119–40, 131, https://doi.org/10.1080/09644016.2020.1827608.

purposeful delineation of how we want to relate to these beings, in the absence of fundamental and unequivocal and unanimously accepted knowledge that would also interpret itself in the same way for all'.[80]

I think that doing justice to this challenge requires a system of representation where multiple voices representing multiple positions or discourses have a place.[81] In the case of Antarctica, maybe the Antarctic Treaty should give space in the deliberation process to the positions discussed in this chapter, but also to more holistic ones, as I discuss next. The Committee for Environmental Protection, I suggest, has already as its mission the protection of Antarctic nature. If its membership were expanded to include representatives of Antarctica itself, and if it grew from being an advisory body to one with decision-making power, the Committee could even become the closest to being Antarctica's representative beyond the Antarctic Treaty—for example, at the Conference of the Parties of the Convention on Biodiversity or at the United Nations Framework Convention on Climate Change (UNFCCC).

5.7 Concluding remarks

Starting from the idea that Antarctica's intrinsic value may reside in its sentient inhabitants, and that this requires representing them in Antarctic deliberation, I claimed in this chapter that independent representation seems more fitting than internalization or encapsulation of interests. Having said that, the challenge is to agree on what should count as their relevant interests. I have shown how the communitarian approach favours the autonomy of wild communities, while the sentientist approach prioritizes the well-being of individual animals. It is thus not obvious which position among them should end up representing non-human animals, and it is not obvious that there will ever be evidence enough to settle the disagreement. This suggests that a plurality of representatives of different positions or discourses might be better suited for the purpose.

Having said this, the claim that the intrinsic value of Antarctica boils down only to the value of its sentient beings sounds like an unnecessary and arbitrary

[80] Mihnea Tanasescu, 'Rethinking Representation: The Challenge of Non-Humans', *Australian Journal of Political Science* 49, no. 1 (2014): 40–53, 49, https://doi.org/10.1080/10361146.2013.874404.

[81] On the representation of discourses, see John S. Dryzek and Simon Niemayer, 'Discursive Representation', *American Political Science Review* 102, no. 4 (2008): 481–93, https://doi.org/DOI:%252010.1017/S0003055408080325. Another alternative would be to have an ombudsman for Antarctic animals. However, ombudsmen do not normally have any decision-making power (Magaña, 'Political Representation', 673).

limitation. Especially given the close dependence of sentient beings on their surroundings, maybe we should also locate the intrinsic value of Antarctica in its existence as one thriving community of life. In the next chapter, I discuss what this position would entail, how it connects to discourses of rights of nature globally, and how it could begin to be implemented.

6
Nothing About Antarctica Without Antarctica

> *You can boil them, bake them, deep-freeze them, crush them, dehydrate them, or even blast them into space. It doesn't matter—tardigrades can survive practically anything. These eight-legged aquatic animals may be small, but they're nearly indestructible.*[1]

This is how one of the coolest Antarctic animals gets portrayed in *National Geographic Kids*. Indeed, water bears—scientifically known as tardigrades—are smaller than a poppy seed and invisible to the human eye, but they are among the toughest creatures on earth. They endure cold temperatures that no other living being could (–200 degrees Celsius in the laboratory) and can survive in a vacuum for prolonged periods of time. They do this by going into cryptobiosis, a state of dormancy where they stop all functions that define life, from breathing to eating, from growing to digesting. Water bears are among the most amazing extremophiles that dare to populate Antarctica (they can also be found on the peaks of the Himalayas and in the depths of the oceans). Found on the coast of the Ross Sea and in Ellsworth Land, these tiny animals are a treat for scientists, but also a potential goldmine for those eager to turn the secrets of their evolutionary success into commercially profitable products.[2]

The most repeated description of Antarctica paints it as the coldest, windiest, driest, highest, and most inaccessible continent on earth. As a result, continental Antarctica is the least floristically and faunistically diverse place on

[1] Kay Boatner, 'Tardigrade', *National Geographic Kids*, https://kids.nationalgeographic.com/animals/invertebrates/facts/tardigrade.

[2] Sarah Zielinski, 'New Water Bear Species Found in Antarctica', *ScienceNews*, 13 July 2014, https://www.sciencenews.org/article/new-water-bear-species-found-antarctica. Extremophiles (of which water bears are just one type) are among the most sought after creatures by those engaged in bioprospecting in Antarctica. Their antifreeze enzymes have potential uses in the agricultural, medical, chemical, and biotechnological industries: Pinki Saini and Pragya Mishra, 'Bioprospecting of Extremophiles for Industrial Enzymes', in *Bioprospecting of Microbial Resources for Agriculture, Environment and Bio-Chemical Industry*, ed. Pinki Saini and Pragya Mishra (Springer Nature Switzerland, 2024), https://doi.org/10.1007/978-3-031-63844-2_3.

From Sovereignty to Guardianship. Alejandra Mancilla, Oxford University Press. © Alejandra Mancilla 2026.
DOI: 10.1093/9780197902769.003.0006

the planet: the only land animals are microscopic mites and wingless insects hiding in moss and lichen. Not surprisingly, most life happens along the coast and in the surrounding seas. Apart from the marine mammals, birds, and fish mentioned in the last chapter, the seabed teams with sea spiders, worms, and crabs. And although it is the only tree-free and shrub-free continent on earth, there are 100 species of moss, 300 to 400 species of lichen, 25 species of liverworts, and 1000 species of fungi in the coastal areas.

Not precisely a biodiversity hotspot, Antarctica is however a hotspot if the criterion used is uniqueness or extremeness, rather than numbers and variety. The very fact that they survive those conditions means that Antarctic inhabitants possess properties that very few species on earth do. Moreover, taken together they constitute an ecosystem of its own, and the one least actively intervened in by humans so far.[3]

In the last chapter, I referred to the Environmental Protocol's claim that the intrinsic value of Antarctica shall be a fundamental consideration 'in the planning and conduct of all activities in the Antarctic Treaty Area'.[4] I suggested that one reading was to locate that intrinsic value in sentient beings. In this chapter, I discuss a different interpretation, namely, one that assigns intrinsic value to Antarctica understood as a self-sustaining ecological community, comprising all its living and non-living entities, sentient or not, including water bears and ice sheets. It could be argued that this is how those drafting the Environmental Protocol imagined that the lines above would get interpreted: if one revises the records of the preparatory meetings, it was ecological rather than animal rights considerations that animated them.

I begin by examining some examples from around the world where natural entities—from rivers to ecosystems—have been granted legal standing, based on their being autonomous, self-functioning entities. Advocates of a growing rights of nature movement have assigned spokespersons or guardians to these entities, and have shaken methodological anthropocentrism, which considers only humans as political and legal subjects. This should give one optimism

[3] I add 'actively' as a qualifier given that indirect human intervention in Antarctica is ubiquitous when it comes to, for example, the presence of nanoplastics and persistent organic pollutants. See, respectively, C. Manno, V. L. Peck, I. Corsi, and E. Bergami, 'Under Pressure: Nanoplastics as a Further Stressor for Sub-Antarctic Pteropods Already Tackling Ocean Acidification', *Marine Pollution Bulletin* 174 (2022): 113176, https://doi.org/10.1016/j.marpolbul.2021.113176; E. Bergami, C. Manno, S. Cappello, M. L. Vannuccini, and I. Corsi, 'Nanoplastics Affect Moulting and Faecal Pellet Sinking in Antarctic Krill (*Euphausia superba*) Juveniles', *Environment International* 143 (2020): 105999, https://doi.org/10.1016/j.envint.2020.105999; and Marco Vecchiato, Elena Argiriadis, Stefano Zambon et al., 'Persistent Organic Pollutants (POPs) in Antarctica: Occurrence in Continental and Coastal Surface Snow', *Microchemical Journal* 119 (2015): 75–82, https://doi.org/10.1016/j.microc.2014.10.010.

[4] Antarctic Treaty System, 'Protocol on Environmental Protection (1991)', Article 3.

when looking at Antarctica: if there is one place on earth where it seems obvious that humans are not part of the picture, this is it. I note, however, that there are political, legal, and institutional challenges that need to be addressed, some of which highlight the uniqueness of the Antarctic case. At the same time, the promise of representing Antarctica is the promise of moving beyond methodological anthropocentrism and nationalism at the global level. I conclude by noting that, if we ultimately aim for a system of political representation for Antarctica based on the *All Affected Interests Principle*, then an institutional change is needed to give voice and decision-making power to entities that have so far been considered as part of the background rather than as subjects on their own right. The absence of easy answers should not detract us from imagining alternative futures, where the political and legal representation of nature becomes the default rather than the exception. Antarctica thus becomes a laboratory for global democratic theory and practice.

6.1 Giving nature rights

In 2008, Ecuador became the first country to give Nature (or rather *Pacha Mama*, Mother Earth) constitutional standing.[5] In the constitution that came into force that year, a whole chapter was devoted to the rights of nature, including 'the right to integral respect for its existence and for the maintenance and regeneration of its life cycles, structure, functions and evolutionary processes' and 'the right to be restored'.[6] Moreover, 'all persons, communities, peoples and nations [could] call upon public authorities to enforce the rights of nature', and they would have 'the right to benefit from the environment and the natural wealth enabling them to enjoy the good way of living'.[7] Two years later, in 2010, neighbouring Bolivia approved the Law on the Rights of Mother Earth, which included among others:

> *the right to maintain the integrity of living systems and the natural processes that sustain them . . . the right to preserve the variety of beings that make up Mother Earth . . . the right to maintain or restore the interrelationship, interdependence, complementarity, and functionality of the components of Mother Earth in a balanced way for the continuation of their cycles and reproduction*

[5] *Constitución de la República del Ecuador.*
[6] Ibid., Articles 71 and 72.
[7] Ibid., Article 74.

of their vital processes . . . the right to timely and effective restoration of living systems directly or indirectly affected by human activities . . . [and] the right to preserve Mother Earth's components from contamination.[8]

This ambitious list was further complemented by the 2012 Framework Law on Mother Earth and Holistic Development for Living Well, in line with the aim of restructuring Bolivia's economic and social system to be in harmony with ecology.[9]

Since then, a lot has been written about whether Ecuador's and Bolivia's ecologically informed laws have really made a tangible difference in the way in which nature is considered when conflicts of interests arise.[10] Given that the economies of both countries still rely mainly on the extraction of natural resources, it has been pointed out that, in practice, the moral, legal, and political mindset is still not in tune with these revolutionary laws; that human interests (narrowly understood as those of the *homo economicus*) have largely trumped the interests of *Pacha Mama*, and that the implementation challenges are daunting.[11] Indeed, the change cannot happen overnight, and there are powerful economic interests against such laws, yet some are optimistic that their very existence will lead humans to see our relation with nature in a new light. As Andreas Gutmann points out:

there is no fixed status of 'cosmic justice' . . . or harmony encompassing the entire society, which can be reached by balancing the rights of all community members and will last forever. The sumac kawsay [living well] is an ongoing process rather than a utopia.[12]

In both Ecuador and Bolivia, the main proponents of rights of nature were Indigenous groups and small farmer organizations, who saw this language

[8] David R. Boyd, *The Rights of Nature: A Legal Revolution that Could Save the World* (ECW Press, 2017), 192–93.

[9] Ibid., 194–95.

[10] See, for example, Louis J. Kotzé and Paola Villavicencio Calzadilla, 'Somewhere Between Rhetoric and Reality: Environmental Constitutionalism and the Rights of Nature in Ecuador', *Transnational Environmental Law* 6, no. 3 (2017): 401–33, https://heinonline.org/HOL/P?h = hein.journals/tevl6&i = 421, and Paola Villavicencio Calzadilla and Louis J. Kotzé, 'Living in Harmony with Nature? A Critical Appraisal of the Rights of Mother Earth in Bolivia', *Transnational Environmental Law* 7, no. 3 (2018): 397–424, https://doi.org/10.1017/S2047102518000201.

[11] In the most prominent case so far, where the rights of a river prevailed over the right to build a road, the argument used to defend the rights of the river to not become a sewage dump was that doing so would damage people: Andreas Gutmann, 'Pachamama as a Legal Person? Rights of Nature and Indigenous Thought in Ecuador', in *Rights of Nature: A Re-Examination*, ed. Daniel P. Corrigan and Markku Oksanen (Routledge, 2021), 44.

[12] Ibid., 45.

as the most appropriate way to translate their own values into the standard Western legal framework. Underlying them is a worldview where humans are in a horizontal and egalitarian rather than a vertical and hierarchical relationship with nature, and where both should flourish in harmony. In the words of one of the members of the Ecuadorian Constitutional Assembly, Sofía Espín: 'Because we are part of nature, to speak of nature's rights is to speak of the rights of communities, the rights of the Huaorani, the rights of the uncontacted tribes, because they live from nature'.[13] Recognizing the rights of nature was thus a recognition of their own, human rights.

A few years later, on the other side of the Pacific, Aoteoroa/New Zealand came up with two pieces of groundbreaking law. In 2014, Te Urewera National Park (government-owned until then) gained legal personhood. Since then, it has no longer been managed by the Department of Conservation, but rather represented by a board of trustees mandated to act 'on behalf of, and in the name of [it]'.[14] The board is formed of both Tūhoe and government appointees.[15] The Tūhoe, who have historically inhabited the land, also provide the guiding principles for the board's decision-making: '*mana me mauri* (the sensitive perception of a living and spiritual force in a place), *rahui* (the prohibition or limitation of a use for an appropriate reason), and *tapu* (a state or condition that requires certain respectful human conduct, including raising awareness or knowledge of the spiritual qualities requiring respect)'.[16] In 2017, the Whanganui River (the third longest in the country) also acquired legal personhood and had two individuals appointed to serve as its guardians. As explained by the environmental lawyer David R. Boyd:

> *the Whanganui River is no longer owned by humans but by itself, Te Awa Tupua. The river and surrounding ecosystem, plus associated metaphysical elements, have legal rights, standing, and an independent voice. The innate values, interests, and status of the river are to be explicitly considered, respected, and upheld. This is consistent with the Māori vision that rivers and other elements of nature have intrinsic value and are inherently incapable of being owned in the western sense.*[17]

[13] Boyd, *Rights of Nature*, 170.
[14] Ibid., 153.
[15] Governing Board of Te Uru Taumatua, 'Governance of Te Urewera', 2025, https://www.ngaituhoe.iwi.nz/te-urewera-governance.
[16] Boyd, *Rights of Nature*, 153–54.
[17] Ibid., 140–41.

These developments and others have made a growing number of environmental activists and theorists hopeful that the methodological anthropocentrism that has characterized Western and (by inheritance) international law will eventually be left behind. Since Christopher Stone's seminal article on whether trees should have legal standing, there have been several court cases demanding the recognition of rights for apes, elephants, endangered species, and unique ecosystems—among others.[18] There have also been attempts to demand rights for more encompassing entities—not least the earth itself. The Universal Declaration of the Rights of Mother Earth, which emerged from the 2010 People's Congress on Climate Change and the Rights of Mother Earth, triggered a discussion at the United Nations General Assembly that has been ongoing since 2011.[19] Although there have been no binding documents so far, the initiative has led to consider our relationship with nature in a global forum, something that would have been unthinkable a few decades ago.

With this in the background, it is a natural follow-up to consider Antarctica as having rights. This has inspired a group of activists and scholars to promote a vision of Antarctica as an independent entity worthy of moral, legal, and political consideration. Because it is under development as I write these lines, in what follows I examine the main foundations of the proposal as presented by one of its leading advocates, Cormac Cullinan, and then turn to the challenges as well as the promises it raises.[20]

6.2 Giving rights to Antarctica

The environmental lawyer Cormac Cullinan is one of the main figures behind the Declaration of the Rights of the Antarctic (hereafter, Antarctica

[18] Stone, *Should Trees Have Standing?*. Other examples where rights have been given to natural entities are the Atrato River, in Colombia, and the Mar Menor lagoon, in Spain. See, respectively Philipp Wesche, 'Rights of Nature in Practice: A Case Study on the Impacts of the Colombian Atrato River Decision', *Journal of Environmental Law* 33, no. 3 (2021): 531–55, https://doi.org/10.1093/jel/eqab021 and Erik Stokstad, 'This Lagoon Is Effectively a Person, Says Spanish Law that's Attempting to Save It', *Science*, 29 September 2022, https://www.science.org/content/article/lagoon-effectively-person-says-spanish-law-s-attempting-save-it.

[19] 'Universal Declaration for the Rights of Mother Earth', Global Alliance for the Rights of Nature, 22 April 2010, https://www.garn.org/universal-declaration-for-the-rights-of-mother-earth/.

[20] Antarctic Rights, 'A Voice for Antarctica'. As I explained in Chapter 1, I am part of the group on a fully voluntary basis. I feel very grateful for that, and I don't think that my views on the present and future of Antarctica would be the same had I not been part of our lively discussions. Having said that, what I develop in this chapter is my own view and does not necessarily represent the group's vision. What unites us, rather, is the realization that there is an obvious normative gap in the governance of Antarctica and other similar places, and it is time to fill it.

Declaration), a cosmopolitan civil society initiative that advocates for the legal rights of the continent as a self-regulating and autonomous realm.

Inspired by theologian Thomas Berry's *Earth Jurisprudence*, Cullinan conceptualizes Antarctica as a

> *magnificent, wild, self-regulating and awe-inspiring living community of inter-related ecological beings . . . The simple act of conceiving of Antarctica and the members of that community as subjects ('beings' or 'entities') with their own interests rather than merely as objects which humans act upon disrupts how states, and many people, relate to Antarctica, and by implication the Antarctic Treaty System.*[21]

Cullinan rightly notes that Antarctica's current governance is deficient, since none of the parties 'are entitled to speak for Antarctica and its non-human inhabitants, nor do they have a mandate to represent its best interests as they do for themselves'.[22] Consequently, Antarctica should be accorded a status akin to a sovereign state, 'to be represented in international and national decision-making processes and courts'.[23]

Echoing the Universal Declaration of Rights of Mother Earth (of which he was the leading proponent), Cullinan outlines a series of principles that should guide our future interactions with Antarctica. Among them: 'Antarctica is an ecological community of inter-related beings', a 'self-regulating and autonomous entity' with 'the rights to exist and the freedom to be wild' and the right 'to a voice in human decision making' such that 'activities contrary to her interests should not be allowed'.[24]

Cullinan's philosophical and legal commitments towards nature are developed in detail in his book *Wild Law*. Contrary to appearances, he claims, the title is not an oxymoron, but rather the direction towards which our legal systems should move. As humans, we have a duty to look at 'the fundamental laws and principles of the universe, since these provide the ultimate framework within which any human legal framework must exist'.[25] In other words, natural, physical laws are to guide and shape human laws, not just in the sense of delimiting the realm of the possible, but also in the sense of

[21] Cullinan, 'Antarctic Realm'.
[22] Ibid.
[23] Ibid.
[24] Ibid.
[25] Cormac Cullinan, *Wild Law: A Manifesto for Earth Justice*, second edition (Green Books, 2011), 78.

showing what is good and valuable. Moving from *is* to *ought* is what ought to be.[26]

Drawing from Berry, Cullinan claims that the 'the first principle of earth jurisprudence must be to give precedence to the survival, health and prosperity of the whole Community [i.e. the earth] over the interests of any individual or human society'.[27] This sounds like a controversial principle to hold, but Cullinan adds that it is complemented with three rights that every individual (human and non-human) has: 'the right to be, the right to habitat, and the right to fulfil its role in the ever-renewing processes of the Earth Community'.[28] According to Cullinan, there is therefore no contradiction between the rights of the whole and the rights of its parts, because the latter are subsumed under the former. Zebras, for example, have no right not to be eaten by lions, because 'part of the natural role of zebras is to provide nourishment for lions'.[29] In being eaten, the zebra 'fulfils the role of zebrakind' and, in so doing, contributes to the good of the whole system.[30]

We should judge actions, following *Wild Law*, according to whether or not they increase the integrity and health of the whole system and the quality and intimacy of the relationships between the component parts (note here the similarities with Aldo Leopold's land ethic, where what is right is to preserve the integrity, beauty, and stability of the biotic community, and what is wrong is to do otherwise[31]). The trick, of course, is to know what is whole-making and what is whole-shaking. Cullinan admits that this is not easy, but recommends appealing to sympathetic introspection, science, and traditional wisdom to find answers.

Cullinan is not shy about his ecocentric approach to law and morality. This position raises many philosophical questions, from the realm of ontology (how to delimit wholes and parts?) to epistemology (how to know nature's interests?) to ethics (how to weigh the interests of wholes vis-à-vis the interests of their parts?). Last, but not least, there is the question of implementing this vision in the real world. In what follows, I will disappoint the philosophers by leaving aside these lines of inquiry. There is abundant literature arguing for and against the possibility of individualizing natural wholes; discussing the coherence (or

[26] The environmental philosopher Holmes Rolston III expressed a similar sentiment after his visit to Antarctica: 'Here what is, ought to be!': Holmes Rolston III, 'Environmental Ethics on Antarctic Ice', *Polar Record* 36, no. 199 (2000): 289–90, 290, doi:10.1017/S0032247400016764.

[27] Cullinan, *Wild Law*, 100.

[28] Ibid., 101.

[29] Ibid., 106.

[30] Ibid.

[31] Aldo Leopold, *A Sand County Almanac, and Sketches Here and There*, special commemorative edition (Oxford University Press, 1989), 224–25.

not) of referring to nature's 'good' or 'purpose'; and examining the difficulty of balancing the interests of individual humans, non-humans, and natural wholes when these come into conflict (assuming that it is meaningful to talk about the interests of entities that are not human).[32] Instead, I take a more pragmatic approach that rescues what seem to me to be the two most important insights of Cullinan's proposal. First, by rejecting methodological anthropocentrism, it provides a fair diagnosis of what has gone wrong in norming our relationships with nature in general, and Antarctica in particular. Second, by rejecting methodological nationalism, it reminds us of the limitations of our current global political system, where individual states are—at least formally—the power-wielders par excellence, even in cases where successful territorial governance requires ignoring national borders.

I thus assume that we can agree on a shared understanding of Antarctica as one distinct ecosystem, located south of the Southern Ocean Convergence. I also assume that it is possible to agree, if not on what constitutes the *summum bonum* of Antarctica, at least on what would be its *summum malum*, namely, what would be damaging for it (e.g. its partial or total destruction). Finally, I grant that—apart from its ill reputation when applied to human communities—giving rights to wholes vis-à-vis individuals has the potential to create conflicts of interests that will need to be resolved through criteria that are probably different from the ones we use for resolving conflicts of interests between individuals only. However, I do not take this as evidence that giving rights to nature is meaningless. Rather, I take this as a necessary implication of acknowledging the complexity of enlarging the sphere of moral, political, and legal consideration.

6.3 Political challenges

If one looks at the movement of rights of nature globally, one obvious feature is that there tends to be a well-delineated local group that stands behind the efforts. Not surprisingly, part of the motivation for the fight is to protect the place where they live and to manage it according to their own principles—which many times diverge from those of the territorial authority under which they fall. As mentioned before, the demand for rights of nature in many of these cases can be interpreted simultaneously as a demand for the group rights of

[32] For discussion of these questions, see Daniel P. Corrigan and Markku Oksanen, eds., *Rights of Nature: A Re-Examination* (Routledge, 2021), https://doi.org/10.4324/9780367479589.

those nature defenders to be respected, and for historical injustices against them to be corrected.[33] But where would such political momentum for the rights of Antarctica come from, absent any human Antarcticans?

This could be called the *feasibility* challenge. Feasibility has become a buzzword among political philosophers, who use it to refer to the likelihood of a proposal being realizable or not.[34] In this case, those worried about feasibility could point out that, given how Antarctic governance works, it is highly unlikely to expect that its main political actors—namely, states—will make any move to incorporate the representation of Antarctica itself within the system. It seems, thus, that rights for Antarctica are an unfeasible proposal, with the slimmest chance of being implemented in the real world. The Antarctica Declaration may have symbolic and expressive value, but it is politically unrealistic, especially in the times we are currently living.

As critics of this notion have pointed out, however, feasibility can become an excuse for justifying the status quo and preventing important and much needed changes from happening.[35] At the extreme, what is feasible just means what is most convenient for those in power. Instead, if one assumes even a moderate dose of political goodwill, creativity, and openness to alternatives, the scope of the feasible expands in ways that may even allow—in this specific case—the representation of Antarctica. To this, one may add the importance of what Pablo Gilabert calls 'dynamic duties', the role of which is not to attain certain immediate outcomes given the circumstances, but to change those circumstances so that certain desirable outcomes become realizable in the middle and long term.[36]

At the domestic level, nature advocates fight for the recognition of nature within the political constituency, 'bringing into public view particular ecological concerns, values, beings, entities, and interests that the advocate believes have been unfairly marginalized, "misrecognized" or "misrepresented" in political discourse and practice'.[37] Here, analogously, it could be up to the 'greener' states within the Antarctic Treaty to start the discussion. Just like in 1989 Australia and France rejected the Mineral Convention and proposed an

[33] See, for example, Catherine J. Iorns Magallanes, 'Nature as an Ancestor: Two Examples of Legal Personality for Nature in New Zealand', *VertigO—La revue électronique en sciences de l'environnement*, hors-série 22 (2015), https://doi.org/10.4000/vertigo.16199.

[34] For an overview, see Nicholas Southwood, 'The Feasibility Issue', *Philosophy Compass* 13, no. 8 (2018): e12509, https://doi.org/10.1111/phc3.12509.

[35] Hallvard Sandven, 'Feasibility: Superfluous or Moralized', *Ethics* 136, no. 1 (2025): 88–120. https://doi.org/10.1086/736744.

[36] Gilabert, 'Justice and Feasibility'.

[37] Robyn Eckersley, 'Representing Nature', in *The Future of Representative Democracy*, ed. John Keane, Sonia Alonso, and Wolfgang Merkel (Cambridge University Press, 2011), 236.

Environmental Protocol instead, here those leading the way could be, for example, states that have underlined the need to consider the future of Antarctica during global climate negotiations.[38] In this, they could be backed up by nongovernmental groups such as Antarctic Rights and other environmental NGOs. Even if this happened, however, there would still be the question of how to choose the actual representatives. The difference between rights of nature movements elsewhere and Antarctic Rights appears again: in the former case, local communities have acquired the role of partial or sole guardians, while here there is no local community to appeal to.

It might be worth inquiring, then, why it is thought that local communities are the most appropriate representatives for natural places (as different domestic constituencies around the world have done). I suggest that there are three main reasons: first, they are relevantly affected by decisions that concern the governance of the place; second, they know the place; and third, they care about it. Local communities are thus a good proxy for picking representatives for the natural places they inhabit, rather than there being something in principle special about them being local.[39] Following from this, one could say that representatives for Antarctica (or for any other natural entity or place not obviously connected to a group of humans) should be *those who know* and *those who care*.

6.3.1 Those who know 1.0

Cullinan explicitly suggests that, in the case of Antarctica, scientists are in a privileged position to become its representatives; more specifically, those who study 'how this community functions and . . . is affected by climate change and other anthropogenic impacts'.[40]

The idea that scientists are potentially the best spokespersons for the nonhuman natural world is not new. Asking how best to represent nature in deliberative democracies, John O'Neill points out that the factors that contribute to give legitimacy to human representation are absent in the case of nature

[38] See, for example, United Kingdom, Australia, Belgium et al., 'Antarctica in a Changing Climate', WP32, ATCM 16, ATCM XLIII – CEP XXIII, 29 April 2021, https://documents.ats.aq/ATCM43/wp/ATCM43_wp032_e.docx.

[39] Some would mention a fourth reason here; to wit, their historical attachment and, arguably, historical rights to the place. But this should be neither a necessary nor a sufficient condition for guardianship: many of those invoking historic rights over a territory have a poor record of protecting it and have not shown the kind of care that is relevant in this context.

[40] Cullinan, 'Antarctic Realm'.

representation. These factors are the presence of actual representatives from different groups, their accountability towards the represented, and their authorization by the represented. Lacking these, O'Neill proposes,

> *the remaining source of legitimacy is epistemic. Those who claim to speak on behalf of those without voice do so by appeal to their having knowledge of the objective interests of those groups, often combined with special care for them.*[41]

O'Neill goes on to suggest that scientists are privileged candidates, but that this strategy is not trouble-free. First, as I showed in the last chapter and as it should become clear in this one, deciding 'which knowledge claims are relevant for the representation of nature', as O'Neill puts it, is not uncontroversial and will not always yield the same prescriptions.[42] In the case of Antarctic science, scientists do not form a monolithic, cohesive front. Depending on their expertise and even within the expert groups (ecology, biology, geology, glaciology, earth systems science, and so on), they have different and even contradictory views. Anyone who has ever been to the biannual conference of the Scientific Committee on Antarctic Research (SCAR) or, for that matter, anyone who has taken a course on the sociology of scientific knowledge knows how diverse scientists' views are, how much they discuss each other's methodologies and results, and how difficult it is to unite them under one voice. To be clear, I am not saying that it is impossible for scientists to agree on anything: the Antarctic scientific community, for example, does not dispute that there is climate change and that it is affecting Antarctica in dramatic ways.[43] A different thing, however, is to get them to agree on the scale, the levels, and the timeframes within which this will happen, and the most efficient ways in which it could be slowed down.

Second, scientists may tend to give too much weight to the interests of nature, to the detriment of human groups who have been and still are marginalized from politics. O'Neill gives the example of fortress conservation policies that have displaced whole communities for the sake of keeping nature 'intact'.[44]

[41] John O'Neill, 'Representing People, Representing Nature, Representing the World', *Environment and Planning C: Government and Policy* 19, no. 4 (2001): 483–500, 496, my emphasis, https://doi.org/10.1068/c12s.

[42] Ibid.

[43] See, for example, Scientific Committee on Antarctic Research, 'The 1st SCAR Antarctic and Southern Ocean Science Horizon Scan—Final List of Questions', *Nature* (Supplementary Information) 512 (2014): 2–5, https://www.nature.com/news/polopoly_fs/7.18965.1407332749!/suppinfoFile/512023a_s1.pdf.

[44] 'Fortress conservation' refers to a conservation model based on the idea that the best way to preserve biodiversity is to leave nature undisturbed by humans, and even to remove them from nature if

It is interesting to note that, in the case of Antarctica, giving the voice to scientists would probably have the opposite effect: rather than banning themselves from the continent to keep it 'intact', they would probably argue for even more presence (as much as they might agree that tourism and fishing should be limited or even prohibited).

A third, related complication is the assumption that both Cullinan and O'Neill seem to make, that scientists 'often care' in the normatively relevant sense, that is, in the sense of being able to act in an impartial, well-informed, and attentive manner towards the object of consideration. Although it would be great if they did, this is not a straightforward assumption to make. Many scientists, unfortunately, care much more about getting grants and new research projects that will keep allowing them to travel to the continent and publish about it than about the natural world in which they conduct their inquiries. Many of them, that is, take a purely instrumental approach that rights of nature advocates would reject.

6.3.2 Those who care

If scientists are not to be trusted as sole Antarctic representatives, one may also turn, as Cullinan suggests, to those who are good at sympathetic introspection and/or to those possessing traditional wisdom. Robin Eckersley makes a similar point when looking for potential nature advocates, noting that 'authority to represent nature might also derive from traditional, local or vernacular knowledge, or from other forms of knowledge or "moral capital"'.[45] Eckersley has in mind environmental NGOs and groups, environmental activists, poets, artists, and documentary-makers, among others.

If a normatively relevant trait of a representative should be to care for their represented, it seems that the kind of people mentioned by Cullinan as engaging in 'sympathetic introspection' and those mentioned by Eckersley as having 'moral capital' would be just as suitable as the scientists. On this note, it is no coincidence that most participants in the Antarctic Rights working group are not natural scientists, but environmental legal activists, humanities scholars, and members of environmental advocacy groups and organizations.[46]

necessary. For a critical account, see Rosaleen Duffy, *Nature Crime: How We Are Getting Conservation Wrong* (Yale University Press, 2010).

[45] Eckersley, 'Representing Nature', 252.
[46] Antarctic Rights, 'Working Group', 2025, https://antarcticrights.org/about/members/.

People who fit these profiles tend to be more open-minded towards a vision of Antarctica as a 'magnificent, wild, self-regulating and awe-inspiring living community of inter-related ecological beings'.[47] If these were the kinds of human agents representing Antarctica, there would be a more or less general agreement that it is best 'to leave Antarctica alone', following Holmes Rolston III's claim: 'Do not nationalize it. Do not internationalize it. *Naturalize* it!'[48]

Even then, however, there would be more and less demanding views on what 'naturalizing' Antarctica would require. Just to give two examples: when Greenpeace established a permanent base at Cape Evans in the Ross Dependency, in 1987, the goal was to press the Antarctic Treaty members to declare Antarctica as a 'World Park'. By this, they meant banning all commercial exploitation and pollution from the continent and restricting human presence to only limited scientific research.[49] There are others, however, who do not oppose all commercial activity, especially not tourism. When John Muir took President Theodore Roosevelt to Yosemite in 1903, his goal was to show him the magnificence of the landscape and convince him of the need to protect it. The three-night camping tour worked wonders. Back from the trip, Roosevelt expanded federal protection of the park and '[signed] into existence five national parks, 18 national monuments, 55 national bird sanctuaries and wildlife refuges, and 150 national forests'.[50] In a similar spirit, some claim that taking people to Antarctica can be seen as offering them an environmental rite of passage, after which they will go back to their daily lives transformed and will act as its ambassadors. Thus, we should not necessarily see human presence as damaging, but even as necessary for protection.[51] These two very different interpretations of what is in Antarctica's best interests are both held by people that, at least from the outside, seem committed in good faith to its protection.

[47] Cullinan, 'Antarctic Realm'.

[48] Rolston, 'Environmental Ethics', 290, my emphasis.

[49] Leslie Carol Roberts, 'The Gigaton Ice Theatre: Performing Ecoactivism in Antarctica', in *Performing Ice, Performing Landscapes*, ed. Carolyn Philpott, Elizabeth Leane, and Matt Delbridge (Springer International Publishing, 2020), https://doi.org/10.1007/978-3-030-47388-4_9.

[50] National Park Service, 'Roosevelt, Muir, and the Grace of Place', last updated 26 August 2021, https://www.nps.gov/yose/learn/historyculture/roosevelt-muir-and-the-grace-of-place.htm.

[51] The argument of ambassadorship is used a lot by tourism companies, selling not just a trip to Antarctica, but a spiritual transformation. For critical views, see Robert B. Powell, Stephen R. Kellert, and Sam H. Ham, 'Antarctic Tourists: Ambassadors or Consumers?', *Polar Record* 44, no. 3 (2008): 233–41, https://doi.org/10.1017/S0032247408007456; Eke Eijgelaar, Carla Thaper, and Paul Peeters, 'Antarctic Cruise Tourism: The Paradoxes of Ambassadorship, "Last Chance Tourism" and Greenhouse Gas Emissions', *Journal of Sustainable Tourism* 18, no. 3 (2010): 337–54, https://doi.org/10.1080/09669581003653534. Moreover, this argument overlooks the difference between taking normal tourists as potential ambassadors and taking individual dignitaries who have the power to actually change things. I thank Janet Martin-Nielsen for making this observation.

6.3.3 Those who know 2.0

Within the group of nature's advocates, those with Indigenous worldviews are also mentioned as having an important role to play. Beyond Antarctica, this makes a lot of sense, not least because their perspectives have been systematically ignored and including them in decision-making processes is seen as part of a larger process of historical reparation. Moreover, as many environmental philosophers have pointed out, the Western human/non-human dichotomy that is to blame for much of our environmental crisis is absent in many Indigenous cosmologies (recall, for example, the principles proposed by the Tūhoe to frame the relationship with the Whanganui River).[52] The idea that Indigenous groups have generally managed to live in harmony rather than at war with nature has become commonplace among the environmentally minded, as has the idea that Westerners would do well in drawing inspiration from them in order to save us all from ecological catastrophe.[53]

Appealing to 'Indigenous worldviews' as if they formed a monolithic front is just as problematic as appealing to 'Chinese Philosophy' or 'Western Science'. But it is true that there are some Indigenous cosmologies that seem particularly well-suited if what one is after is the political representation of nature. As shown above, the cases of Ecuador and New Zealand are good examples in this regard. In the case of Antarctica, including Indigenous representatives might sound odd at first sight, given the lack of an Indigenous population. However, it makes a lot of sense in terms of providing different viewpoints that may enrich the deliberation process. On this basis, Priscilla M. Wehi and colleagues have argued for the inclusion of Māori perspectives in the shaping of New Zealand's Antarctic policy. For them, applying a Māori lens to Antarctic management and policy would help transform these and be a useful tool for creating fairer socio-environmental futures, adding stewardship values and practices that have been absent so far in the continent's governance.[54] They mention relationality and reciprocity as two central values that the Māori

[52] See also Plumwood, *Feminism* and Val Plumwood, *Environmental Culture: The Ecological Crisis of Reason* (Routledge, 2002).

[53] See, for example, Claudia Sobrevila, *The Role of Indigenous Peoples in Biodiversity Conservation: The Natural but Often Forgotten Partners* (World Bank, 2008), https://documents1.worldbank.org/curated/en/995271468177530126/pdf/443000WP0BOX321onservation01PUBLIC1.pdf, and Somini Sengupta, Catrin Einhorn, and Manuela Andreoni, 'There's a Global Plan to Conserve Nature: Indigenous People Could Lead the Way', *The New York Times*, 23 June 2023, https://www.nytimes.com/2021/03/11/climate/nature-conservation-30-percent.html?searchResultPosition = 1.

[54] Priscilla M. Wehi, Vincent van Uitregt, Nigel J. Scott et al., 'Transforming Antarctic Management and Policy with an Indigenous Māori Lens', *Nature Ecology & Evolution* 5, no. 8 (2021): 1055–59, https://doi.org/10.1038/s41559-021-01466-4.

worldview could contribute. And, although they do not put it in the language of political representation, it is easy to translate their proposal into these terms.

The Māori, however, offer just one worldview among many possible others that could have a say in the representation of Antarctica, and it is an open question as to how many should be chosen, which ones, and on what grounds. Moreover, there is irony in the fact that it is mostly Westerners who are having the discussion of how to represent Antarctica, and it would probably be Westerners selecting which non-Western views are most appropriate to reach the desired end.

To sum up, scientists, environmentally engaged individuals and groups, and holders of specific kinds of traditional wisdom seem suitable candidates for representing Antarctica understood as one independent entity, based on their epistemological and ethical credentials.[55] Each of them raises challenges of their own. Scientists may know a lot about the continent, but they may not necessarily care for Antarctica non-instrumentally, and may agree with the 'Leave her alone' dictum applied to tourists and fishermen and everyone else, but not to them: they may decide that Antarctica should remain the largest open-air laboratory in the world, even if this requires putting the non-basic interests of humans above the basic interests of natural beings and entities. Individuals and groups who care in a normatively relevant way are more likely to accept the view that Antarctica is an independent entity worthy of being represented. It is not obvious, however, that they would agree with scientists or even among themselves regarding how best to represent Antarctica: some would demand stronger measures against human interference than others. Finally, when it comes to Indigenous worldviews, to include them can be seen as a welcome development in terms of diversity, but it is an open question as to which of those worldviews should be picked and why.

Is this a damning conclusion? I think not, and, in providing an answer, I am indebted to Robin Eckersley's reflections on how to represent excluded others (specifically nature) in deliberative democracies. As a critical political ecologist, Eckersley rejects both that we can acquire objective knowledge of the external world and that the external world is a mere social construction. Instead, she takes an intermediate path, whereby what we can aspire to is intersubjective knowledge corroborated by intersubjective verification, recognizing that 'intersubjective understandings of "reality" will always be historically

[55] A fourth alternative, that I do not discuss here, is to include people chosen by lottery. See Alexander A. Guerrero, 'Against Elections: The Lottocratic Alternative', *Philosophy and Public Affairs* 42, no. 2 (2014): 135–78, https://doi.org/10.1111/papa.12029.

and culturally specific, provisional, and potentially always vulnerable to challenge and change'.[56] The lesson that should be drawn from this, applied to the Antarctic case, is that what should be avoided at all costs is ending up with one spokesperson or representative only, or even a handful of them. Nature, and Antarctica, need a plurality of voices representing them or, as suggested in Chapter 5, representing what we think ought to be our relationship with them. This would mean acknowledging rather than brushing off the complexities of standing for an Other who cannot be present, cannot correct us, and cannot authorize us to speak on their behalf. In this, giving representation to Antarctica as a whole is no different than giving representation to its sentient beings: given the diverse positions among those who know them and care for them, the best way to ensure a fair representation is by setting up a system that reflects that diversity as closely as possible. Instead of directly representing Macaroni Penguins or melting ice sheets, maybe we should think of representing *discourses* around them, some of which could get constituted 'by the very fact of their representation'.[57] Indeed, even if it takes a long and winding road to turn Antarctica and its living and non-living entities into independent political subjects, the fact that different discourses for their representation get articulated and publicized would already be a big step towards altering the methodological anthropocentrism and methodological nationalism in which global bodies and institutions are still imbued.

6.4 Legal and institutional challenges

Giving representation to Antarctica or to different discourses for the representation of Antarctica creates a double challenge for international law: it requires the recognition of a non-human, non-state entity as a political subject, and it requires granting representation to it by non-state actors. The alternative, to keep things as they are, is normatively inadequate and practically insufficient. First, it is normatively inadequate because it stays within the boundaries of methodological anthropocentrism, which considers nature almost exclusively as a resource. This makes its protection contingent on narrow human interests, to the point where minor benefits for humans can end up justifying big and sometimes irreversible harms for nature.

[56] Eckersley, *Green State*, 123.
[57] Dryzek and Niemayer, 'Discursive Representation', 485.

Despite the Environmental Protocol's ambitious goals, note, for example, that only 1.5 percent of ice-free Antarctica (where most terrestrial Antarctic life happens) is protected under the network of Antarctic Specially Protected Areas (ASPAs). These areas are particularly threatened by human proximity (mainly through science and tourism) and by the risk of biological invasions.[58] If Antarctica were a country, it would rank sixty-ninth out of eighty-four countries surveyed in terms of their terrestrial biodiversity protection, between Mali and Kazakhstan.[59] The lack of will of the Antarctic Treaty parties to extend protected areas and restrict human activities is an instance of what Kees Bastmeijer has called 'decision-making by non-decision making': absent consensus to decide on an issue (here, the protection of more extensive ice-free areas), implicit decisions are made (here, letting tourist sites and scientific sites continue to expand).[60]

The Marine Convention is another body which makes momentous decisions for Antarctica, as mentioned in Chapter 5. Although set up as an instrument for the conservation of marine life in the Southern Ocean, it has been increasingly interpreted by some countries as a fishing convention, where the question is not whether to extract Antarctic marine life or not, but rather how much.[61] If one bears in mind that all Antarctic fishing is luxury fishing, this is one more instance of privileging trivial human interests over the basic interests of non-humans.[62] Having Antarctica represented in the Marine Convention could help to move it once again towards its original purpose, or to set up even more stringent rules, where preservation is the default and petitions for rational use bear the burden of proof.

Second, keeping things as they are is practically insufficient. The current global political order, which takes methodological nationalism for granted, leaves out of sight the representation of places like Antarctica, which lie beyond any individual state's jurisdiction, but also places that are located among different state jurisdictions, even though they require one unified framework if they are to be successfully governed from an environmental point of view (an

[58] Justine D. Shaw, Aleks Terauds, Martin J. Riddle, Hugh P. Possingham, and Steven L. Chown, 'Antarctica's Protected Areas Are Inadequate, Unrepresentative, and at Risk', *PLOS Biology* 12, no. 6 (2014): 1–5, https://doi.org/10.1371/journal.pbio.1001888.

[59] Ibid., 2.

[60] Kees Bastmeijer, 'The Antarctic Treaty Turning Sixty: Does Aging Affect Decision-Making? The Case of Antarctic Tourism', presentation at the workshop 'On Its 60th Birthday, Is the Antarctic Treaty in Good Health?', organized by the project 'Political Philosophy Looks to Antarctica', University of Oslo, September 2019.

[61] Goldsworthy, 'Consensus Decision-Making'.

[62] Krill is extracted for human health products and animal nutrition, while Patagonian and Antarctic toothfish and mackerel icefish are sold in high-end markets exclusively.

example of the former are the oceans beyond states' jurisdiction; an example of the latter is the Amazon rainforest, which extends over nine different countries). This gap is obvious in, for example, the Conference of the Parties of the UNFCCC, in which key decisions for Antarctica's future are made, but where Antarctica as such is not represented. Noting this, in 2023, for the first time, the Antarctic Treaty parties recognized that this absence might be an issue. The declaration was a follow-up of a working paper presented in 2021 by a group of Antarctic Treaty parties demanding that the national delegates to the UNFCCC be 'aware of the implications for Antarctica, unless action is taken to avoid dangerous climate change'.[63] A watered-down version of that proposal, the Helsinki Declaration, commits all parties to 'substantially increasing our efforts to communicate the global implications of climate change in Antarctica within our own countries and in international forums, and the need to prevent the irreversible changes to Antarctica and consequential implications for the planet'.[64] This boils down to the claim that, when negotiating, Antarctic Treaty parties should be better at communicating to others what Antarctic science has been repeating for years; namely, that Antarctica is one of the places on earth most affected by climate change, and that something needs to be done about it. But compare this commitment to that of any country that is negatively affected by climate change and is represented in this forum.[65] As we saw in the last chapter, even if national delegates were educated to be more sensitive towards the fate of nature and tried to encapsulate its interest in their decisions, they are ultimately delegates for specific countries, which might have interests that conflict with those of Antarctica. It is not difficult to guess, in these situations, which interest will end up being represented.

In what follows, I pick up where I left off at the end of Chapter 5 and provide some ideas as to how the representation of Antarctica and its individual inhabitants could become a reality, thanks to the performance of certain dynamic duties by certain political actors. The obvious first forum, I suggest, could be the Antarctic Treaty itself, and especially its Committee for Environmental Protection.

[63] United Kingdom et al., 'Antarctica in a Changing Climate'.

[64] Antarctic Treaty Meeting, 'Helsinki Declaration on Climate Change and the Antarctic, Resolution 2', ATCM XLV - CEP XXV, Helsinki, 2023, https://documents.ats.aq/ATCM45/ad/atcm45_ad006_e.pdf.

[65] See, for example, Alliance of Small Island States (AOSIS), 'Alliance of Small Island States (AOSIS) Declaration on Climate Change 2009', United Nations Sustainable Development, 21 September 2009, https://sustainabledevelopment.un.org/content/documents/1566AOSISSummitDeclarationSept21FINAL.pdf.

6.5 The promise of representing Antarctica

In his thought-provoking book *A Blue New Deal*, Chris Armstrong proposes a World Ocean Authority to overlook and coordinate globally what we should and should not do regarding the many challenges that the oceans are facing today, from overfishing and plastic pollution to the destruction of coral reefs and mangrove forests. In this new politics for the oceans, not just humans but also non-humans—especially cetaceans—are foregrounded.[66] Armstrong does not mention representation as the political mechanism to achieve this, but the idea of instituting an independent body in charge of ocean governance aims in part at filling that gap. While today the oceans are absent as a subject of consideration in international fora, Armstrong's World Ocean Authority would stand for them, independently of individual states and other institutions.[67]

I have argued so far that while states' interests are represented in Antarctic governance, Antarctica itself lacks any direct representation. As opposed to the oceans, however, within the Antarctic Treaty, there is already an institutional body that could give space for such representation. The Committee for Environmental Protection, established through the Environmental Protocol, is mandated to 'provide advice and formulate recommendations to the Parties in connection with the implementation of [the] Protocol'.[68] The Committee gives 'specialist environmental, scientific and technical advice'[69] and is composed of delegates from all those countries that have ratified the Protocol (to this date, twenty-nine consultative parties and thirteen non-consultative parties).[70] Importantly, the Committee is joined in its annual meetings (which happen in conjunction with the Antarctic Treaty Meetings) by a number of expert organizations, such as the Scientific Committee for Antarctic Research, the United Nations Environmental Program, and the World Meteorological Organization. As official observers, they submit working papers and information papers and participate in the discussion. The result of the Committee's work is then presented to the Antarctic Treaty Meeting which, based on its recommendations, may adopt measures, decisions, and resolutions.

[66] Chris Armstrong, *A Blue New Deal: Why We Need a New Politics for the Ocean* (Yale University Press, 2022).

[67] See, especially, ibid., chapter 10.

[68] Antarctic Treaty System, 'Protocol on Environmental Protection (1991)', Article 12.

[69] Ewan McIvor and Rodolfo Andrés Sánchez, 'The Antarctic Committee for Environmental Protection: Past, Present, and Future', *Polar Record* 43, no. 3 (2007): 239–46, https://doi.org/DOI: 10.1017/S0032247407006547.

[70] Secretariat of the Antarctic Treaty, 'Parties', 2025, https://www.ats.aq/devAS/Parties?lang = e.

In the context of our discussion, the Committee is a deliberative body where scientists and policy-makers, informed by the best available science, propose what they consider to be necessary policies for Antarctic protection. They are, in a way, already representing Antarctica's interests—although still tied to the standpoints of their individual countries. Yet, the Committee could be the institutional structure through which extending representation to Antarctica happens. An initial step could be to incorporate representatives of alternative knowledge systems, civil society groups, and nongovernmental organizations as observers who are also allowed to contribute material for discussion. Eventually, the Committee could be given decision-making power within the Antarctic Treaty, as the body directly representing Antarctica, and could participate in other inter-state institutions the decisions of which matter to Antarctica: the Antarctic Marine Convention, the Convention on Biodiversity, the UNFCCC, the Biodiversity Beyond National Jurisdiction Agreement, and the United Nations Convention on the Law of the Sea.

To be sure, this requires a considerable leap of political imagination. There are a number of questions regarding implementation that would need to be answered, such as how many representatives Antarctica should have vis-à-vis national delegates, who would fund them, who would hold them accountable (to avoid corruption and human self-interest getting in the way), and how conflicts between human, non-human, and states' interests would get resolved. Furthermore, such a leap requires the existing institutional apparatus to open itself to the possibility that not just humans, but also nature, should be part of the Antarctic demos, leaving behind methodological anthropocentrism. It also requires leaving behind methodological nationalism, insofar as the representatives of Antarctica and its individual inhabitants would not be states and yet would have decision-making power within an international body.

Again, those worried about feasibility may be wondering why states would give up power voluntarily to non-state actors who do not have the leverage to demand it. My answer is that this scepticism is rooted in a realist conception of states as selfish actors striving to thrive in an anarchic setting. If this is replaced by a normative conception of states as fiduciaries of humanity and, eventually, of nature, these possibilities do not look so remote anymore.[71] If it were civil society itself which demanded that nature be considered politically and legally

[71] For states as fiduciaries of humanity, see Eyal Benvenisti, 'Sovereigns as Trustees of Humanity: On the Accountability of States to Foreign Stakeholders', *American Journal of International Law* 107, no. 2 (2013): 295–333, https://doi.org/10.5305/amerjintelaw.107.2.0295 and Evan J. Criddle and Evan Fox-Decent, *Fiduciaries of Humanity: How International Law Constitutes Authority* (Oxford University Press, 2016). For states as fiduciaries of nature, see Mancilla, 'Effective Deoccupation'.

in global institutional arrangements, then—in this view—states should be responsive to that demand.

6.6 Concluding remarks

In the last three chapters, I have examined how the Antarctic political community was initially taken to be all of humanity, represented by a handful of self-appointed states performing substantial scientific research on the continent. I explored what it would mean to represent not just humans, but the actual inhabitants of the territory—in particular, sentient animals. I then proposed a further extension of the Antarctic constituency, where Antarctica as one unit would be the subject of consideration. These extensions would bring Antarctic governance closer to the ideal proposed by the *All Affected Interests Principle*, where all those relevantly affected by a decision ought to be, if not directly deliberators, then at least represented in its making. But there are still other affected parties that have not even been mentioned yet: just as all humanity is considered part of the Antarctic demos, should not all nature (insofar as it may be affected by Antarctic governance) be represented too? And what about all those who still do not exist but will exist in the future, human and non-human: should we not also include them in the decision-making process?

While there are a few theories proposing how to extend representation to future generations (human and non-human) at the domestic and at the global level, a systematic theory for the political representation of nature within and beyond domestic constituencies remains to be written.[72] Antarctica is a perfect laboratory to think through such a theory. Moreover, if the Antarctic Treaty members are daring enough to take steps in that direction, Antarctica could be the first place on earth where it is put into practice. Nothing would be decided about Antarctica without considering Antarctica itself and its non-human inhabitants as political and legal subjects worthy of representation.

[72] An example of a theory for the political representation of future humans and non-humans globally is Stephen M. Gardiner, 'A Call for a Global Constitutional Convention Focused on Future Generations', *Ethics & International Affairs* 28, no. 3 (2014): 299–315. For a theory of how to represent non-human animals in liberal democracies and, to some extent, beyond them, see Cochrane, *Sentientist Politics.*

7
A Model of Joint Guardianship

> *is [the Antarctic Treaty] not an excellent precedent for many other situations in which agreement about activities is required among various countries? Is it not time to think of disseminating to the entire terrestrial globe the principles set forth in this treaty—namely, the close collaboration of countries with different social structures for the exploration of nature and the mastery of a broad region of our earth for peaceful purposes?*[1]

The quote above is from Yevgeny K. Fedorov, chairman of the Soviet Working Group on Rockets and Satellites during the International Geophysical Year (1957–58), head of the Soviet Weather Bureau, committed Communist and, above all, ardent enthusiast of the Antarctic Treaty. For him, the question posed was not rhetorical, but practical. Like other scientists in the 1970s, Fedorov realized that certain challenges would be posed to humanity at large that could not be solved by the actions of individual states: coordination among them would be needed, and the Antarctic Treaty could be the model to follow. Less than two decades later, the Environmental Protocol consolidated Fedorov's view that successful protection of a place like Antarctica required the collaboration of many states. Thanks to it, Antarctica became a 'natural reserve, devoted to peace and science' where human activities were regulated and mineral prospection and extraction were banned, turning Antarctica into—arguably—the most protected territory on earth.[2]

In the previous chapters, I have criticized the Antarctic Treaty for freezing the territorial claims in Antarctica instead of rejecting them altogether; for converting science into political currency, allowing richer states to have decision-making power to the detriment of developing ones, thus preserving global inequalities inherited from the colonial era; for failing to include in the Antarctic demos the

[1] Yevgeny K. Fedorov, 'Antarctica: Experimental Proving Ground for Peaceful Coexistence and International Collaboration', in *Frozen Future*, ed. Richard S. Lewis and Philip M. Smith (Quadrangle Books, 1973), 78–79.

[2] Antarctic Treaty System, 'Protocol on Environmental Protection (1991)'.

From Sovereignty to Guardianship. Alejandra Mancilla, Oxford University Press. © Alejandra Mancilla 2026.
DOI: 10.1093/9780197902769.003.0007

real Antarctic inhabitants, that is, the rich life surrounding the continent; and for recognizing the intrinsic value of Antarctica yet failing to translate this recognition into an active representation of Antarctica in fora like the UNFCCC and the Convention on Biodiversity. In this chapter, following Fedorov, I turn instead to the idea that, despite all the above, one can still praise the Antarctic Treaty as a blueprint for Joint Guardianship, where territorial duties have been emphasized over territorial rights, and where a whole continent has been preserved from resource exploitation, leaving nature—for the most part—alone.

There are a few qualifications to be made before proceeding.

First, what I say in this chapter focuses on what Antarctic Treaty members have done and do *in* Antarctica, which must be distinguished from what they have done and do *for* Antarctica, or *to* Antarctica. This distinction matters because, as I show in Chapter 8, the actions and omissions of these same states beyond Antarctica are harming the continent and putting it at risk.

Second, when referring to the Antarctic Treaty as a model, I look at what the official documents declare to be the objectives of the Antarctic Treaty and its associated legal instruments, rather than at how the parties have interpreted these objectives and implemented measures to achieve them. This is important because, as I showed in Chapters 5 and 6, one could argue that the Environmental Protocol in particular should be read in a much more ambitious light to be truly effective. Moreover, sometimes the rules are there, but there is a laggardness of will among the parties to implement them. To name three examples: ice-free areas (where most terrestrial Antarctic life happens) remain the least protected areas of the continent.[3] When it comes to the most densely inhabited areas, there is also a lot to improve in order to reduce pollution and environmental impact. Anarchy is the best way to describe the organization of energy sources and waste disposal on King George Island's Fildes Peninsula, home to five scientific stations from four different states.[4] Finally, the still unratified annex on liability in case of environmental emergencies suggests that it is much easier for states to be protectors of Antarctica by omission than it is for them to take responsibility for their actions and those of their nationals in Antarctica.[5]

[3] Shaw et al., 'Antarctica's Protected Areas', 2.

[4] Christina Braun, Fritz Hertel, and Hans-Ulrich Peter, 'Environmental Management: The Fildes Peninsula Paradigm', in *Handbook on the Politics of Antarctica*, ed. Klaus Dodds, Alan D. Hemmings, and Peder Roberts (Edward Elgar Publishing, 2017). The stations belong to Chile, China, Russia, and Uruguay.

[5] Alan D. Hemmings, 'Liability Postponed: The Failure to Bring Annex VI of the Madrid Protocol into Force', *The Polar Journal* 8, no. 2 (2018): 315–32, https://doi.org/10.1080/2154896X.2018.1541550; Rüdiger Wolfrum, 'Liability for Environmental Damage in Antarctica: Supplement to the Rules on State Responsibility or a Lost Opportunity?', in *International Law Between Universalism and*

Third, what I refer to as the model of Antarctic governance has not always been intended directly by the parties but has come as a side-effect, or *malgré eux*. As mentioned in Chapter 4, the very birth of the Antarctic Treaty can be seen as an 'agreement to disagree' between territorial claimants who could not recognize each other's claims,[6] and the rejection of the Mineral Convention in favour of the Environmental Protocol was for some of the claimants more of a guarantee that their claims would rest frozen in peace than an active pursuit of environmental protection on the continent.[7] At the same time, when facing the external world, the signatories like to congratulate themselves on these 'achievements' as if they fully owned them.

Fourth, to say that protection has been prioritized over exploitation in Antarctica is not to say that there is no economic exploitation. As has been mentioned before, fishing is the most important economic activity in the Southern Ocean, and it does not fall under the Environmental Protocol's domain, but rather under the Marine Convention.[8] Tourism, while not an extractive activity, still has a direct effect on Antarctic 'living resources', and is growing exponentially, from 950 reported visitors in 1980 to 15,000 in 2000 to 104,000 in 2022–23.[9] Despite repeated talk about the need to draft a specific protocol or at least establish an entry fee for the continent, tourism remains self-regulated under the International Association of Antarctic Tour Operators (IAATO) and is carried out at no extra cost for the companies or the passengers. Lastly, bioprospecting flourishes as an unregulated activity with the parties consistently failing to discuss the issue systematically at the Treaty meetings.

Fifth, there are some aspects of Antarctic governance that need improvement. The rule of consensus, where all parties must agree for measures, decisions, and resolutions to be passed, does not help the system to move faster and to address new challenges, especially as membership grows.

Fragmentation, ed. Isabelle Buffard, James Crawford, Alan Pellet, and Stephan Wittich (Martinus Nijhoff Publishers, 2008), https://doi.org/10.1163/ej.9789004167278.v-1086.264.

[6] James Roberts, 'Flexibility, Ambiguity and the Construction of the Antarctic Treaty System' (Postgraduate Certificate in Antarctic Studies, University of Canterbury, 2011), https://ir.canterbury.ac.nz/server/api/core/bitstreams/caec9094-4f44-48a3-acdc-545bc3b1b390/content.

[7] Auburn, 'Antarctic Minerals', 208.

[8] As mentioned in the last chapter, although the Antarctic Marine Convention is not a fisheries agreement, but rather a convention for the conservation of Antarctic marine living resources, it does allow for 'rational use'—a concept the interpretation of which has created much contestation through the years. For a critique, see Jacquet et al., 'Rational Use'.

[9] For the most recent report, see International Association of Antarctica Tour Operators, 'IAATO Overview of Antarctic Tourism: The 2022–23 Season, and Preliminary Estimates for 2023–24', IP 56, ATCM 17, XLV Antarctic Treaty Consultative Meeting, 2023, https://iaato.org/system/files?file = 2025-01/ATCM45_ip056_e_IAATO-Overview-of-Antarctic-Tourism-The-2022-23-Season-and-Preliminary-Estimates-for-2023-24.pdf.

Sixth and finally, one does not have to be cynical to note that, to the extent that environmental protection has been successful in Antarctica, it is at least partly because Antarctica has protected itself. Too inhospitable, too economically uninteresting, and too distant for states to be bothered, the stakes in protecting it have been low, and the public relations gains have been high: Antarctic Treaty members have been environmental guardians at relatively small cost.

With these qualifications in mind, I examine the claim that Antarctica is the most protected territory on earth thanks to the Antarctic Treaty and, above all, to its Environmental Protocol. I suggest that, without necessarily intending it, its signatories consolidated a unique model of territorial governance: they kept Antarctica as a territory beyond any individual state's jurisdiction, subject instead to the Joint Guardianship of Antarctic Treaty members, and they banned mineral prospection and extraction, effectively focusing on preservation. I go on to suggest that, in times where preserving certain ecosystems of global importance becomes crucial, this model—however imperfect—may serve as inspiration. I address four objections: that applying this model beyond Antarctica would be unfair for states, that it would be unfair for the people living in ecoregions, that it would be unfeasible, and that there are other existing models that would be more suitable. I conclude by reviewing the merits of the model while pointing at the same time to its limits, and to the need for extending the Antarctic demos beyond humans.

7.1 Neither Permanent Sovereignty nor Common Heritage

In Chapter 4, I mentioned how the Antarctic Treaty parties started negotiations at the beginning of the 1980s to regulate the prospection and extraction of mineral resources on the continent. Confronted with a global energy crisis, member states looked to Antarctica as the next, obvious target in the quest for oil and gas, intending to negotiate a Mineral Convention as the legal instrument that would regulate exploitation. Around the same time, another crisis also made Antarctica an object of interest: prolonged droughts and water shortages around the world, from the Middle East to Southern California, made some entrepreneurs and scientists think of harvesting icebergs as a solution. One of them, the Saudi prince Mohammed Al Faisal Al Saud, even sponsored two conferences on the topic, at the University of Iowa (1978) and at the Scott Polar Research Institute in Cambridge (1980). The idea of towing Antarctic ice to dry regions further north was met with everything from dry scepticism ('Is iceberg water a real possibility or just a pipe dream?') to cautious optimism, with some presenters going as far as showing what it would take to construct an iceberg processing plant.[10]

Thanks to the Environmental Protocol, however, neither oil rigs were installed nor ice harvested. Instead, Antarctica became not just the last and only non-militarized, non-nuclearized continent on earth[11] but also the last and only continent under states' governance yet beyond states' exploitation. Since the Antarctic Treaty was signed in 1959, the freezing of territorial claims has meant that member states effectively focused on their territorial duties, guaranteeing peace and scientific cooperation, and mutually observing their behaviour through a system of open inspections.[12] With the signature of the Environmental Protocol, a new commitment was added: namely, to protect Antarctica in its dual role as the largest open-air laboratory for the natural sciences and the largest unspoiled wilderness on earth.

In retrospect, one could say that, even though it was designed more to save face among the original claimants than to contribute a new model to international law, the Antarctic Treaty's freezing of sovereign rights over the continent was an unprecedented and welcome outcome. Moreover, even though the Environmental Protocol was negotiated hastily, it strengthened the Antarctic Treaty's explicit aim of being an international arrangement for the benefit of all humankind and arguably also for the benefit of Antarctica itself: rare among international legal documents, protecting the intrinsic value of the place figured centrally among its objectives.

To understand how this could happen, it is useful to contrast the Antarctic situation with all other areas within and beyond national jurisdiction. In Antarctica, as opposed to these other areas, neither the doctrine of Permanent

[10] See, respectively, David J. DeMarle, 'Design Parameters for a South African Iceberg Power and Water Project', in *Proceedings of the Conference on Use of Icebergs: Scientific and Practical Feasibility, Cambridge, UK, 1–3 April 1980* (International Glaciological Society, 1980) and W. F. Weeks, 'Iceberg Water: An Assessment', in *Proceedings of the Conference on Use of Icebergs: Scientific and Practical Feasibility, Cambridge, UK, 1–3 April 1980* (International Glaciological Society, 1980). The first volume of the journal *Annals of Glaciology* was in fact a compilation of the papers presented at the second conference.

[11] It is the only continent where there is a total prohibition against testing of nuclear weapons and installation of nuclear waste disposal.

[12] Open inspections have been criticized for their tendency to focus on information exchange and capacity building to the neglect of their real aim, which is monitoring that stations are complying with the rules. Most criticisms, however, are not directed against the idea, but rather regard how to improve the way in which inspections are carried out. See, for example, Sizhu Liu, 'Ensuring Compliance with Antarctic Environmental Protection Through Inspections: Problems and Prospects', *Marine Policy* 171 (2025): 106433, https://doi.org/10.1016/j.marpol.2024.106433; Sune Tamm, 'Peace vs. Compliance in Antarctica: Inspections and the Environment', *The Polar Journal* 8, no. 2 (2018): 333–50, https://doi.org/10.1080/2154896X.2018.1541551.

Sovereignty over Natural Resources nor the doctrine of Common Heritage of Mankind applied.

At the state level, territorial rights comprise three different types of rights: rights of jurisdiction, rights to border control, and rights over natural resources.[13] This last type of rights was enshrined through the Doctrine of Permanent Sovereignty over Natural Resources (hereinafter, Permanent Sovereignty), which says that

> *The right of peoples and nations to Permanent Sovereignty over their natural wealth and resources must be exercised in the interest of their national development and of the well-being of the people of the State concerned.*[14]

Adopted as a United Nations General Assembly Resolution in 1962, Permanent Sovereignty was the economic wing of the Declaration on Decolonization. Newly independent countries wanted not just political but also economic self-determination. To achieve it, they needed control over their land and their resources, to be able to use them for their own development. At the time, Permanent Sovereignty seemed fully justifiable. Yet, as much as it was and still is instrumental, especially for developing countries that base their economic growth on extractive economies, Permanent Sovereignty is sorely inadequate from an environmental point of view.[15] This is because the doctrine has been interpreted mainly as the right of individual states to use and exploit, with preservation only as an afterthought; and as a right that states can exercise without any requirements to coordinate with others, unless unavoidable. Here, its defenders might complain that there is nothing in the doctrine itself that necessitates this; on the contrary, they could add, a proper understanding of it includes both rights and duties.[16] They could even point to some nature conservation success stories in states like Costa Rica or Bhutan. My response to this is threefold. First, these cases are more the exception than the rule, especially considering the current trend of ecologically harmful resource and land grabbing at a global scale, many times facilitated by national governments.[17] Second, successful conservation stories are normally grounded in economic

[13] Miller, 'Territorial Rights', 252.

[14] United Nations General Assembly, 'Permanent Sovereignty over Natural Resources'.

[15] For why Permanent Sovereignty is also unjustifiable for other reasons, see Chris Armstrong, 'Against "Permanent Sovereignty" over Natural Resources', *Politics, Philosophy & Economics* 14, no. 2 (2015): 129–51, http://journals.sagepub.com/doi/abs/10.1177/1470594X14523080.

[16] See, for example, Nico Schrijver, *Sovereignty over Natural Resources: Balancing Rights and Duties* (Cambridge University Press, 1997), https://doi.org/http://dx.doi.org/10.1017/CBO9780511560118.

[17] See, for example, Fred Pearce, *The Land Grabbers: The New Fight over Who Owns the Earth* (Beacon Press, 2012).

motives; for example, insofar as they attract tourists, it seems profitable to keep certain natural sites protected. But this seems too precarious a defence for places that we should aim to preserve regardless of how pleasing they are for prospective visitors. Third, making the fate of 'natural resources' dependent upon the governments in office seems inadequate for the purpose: their long-term conservation (or not) should not be reliant upon the short-sighted goals of fleeting politicians. In short, looking at how states have exercised and still exercise Permanent Sovereignty, making preservation rather than exploitation its default understanding would require a major 'conceptual amelioration'.[18]

Because in Antarctica territorial rights were never settled, none of the claimants could invoke this right to exploit the continent, even though this was one of the central motivations behind the claims. At the same time, because Antarctica was not internationalized (the territorial rights of individual states, although in frozen form, nonetheless existed), the doctrine of Common Heritage of Mankind (hereinafter, Common Heritage) could not be applied either.

Like Permanent Sovereignty, Common Heritage crystallized in the 1960s as a demand from developing states to ensure that the few remaining global commons would be shared equitably, rather than appropriated on a first-come, first-served basis.[19] In particular, what was at stake was the international administration of the deep seabed, which was seen as providing a 'noble dream of seabed riches' from which the Third World especially could benefit.[20] Legal scholar Christopher Joyner summed up the logic behind it:

> *the Common Heritage concept, like the regimes for res nullius and res communis, essentially furnishes a legal device for exploiting resources of the commons. Such exploitation becomes fundamentally necessary if that regime is to redistribute global wealth successfully, as the New International Economic Order ideology of the 1970s might have had it. Not surprisingly, then, arguments made for Common Heritage stressed more the immediate political and*

[18] 'Conceptual amelioration' is a term used by philosophers of language as a response to the fact that 'sometimes, our conceptual schema must be ameliorated to attain certain beneficial consequences, which may be social, theoretical, political, or otherwise': Manuel Gustavo Isaac, Steffen Koch, and Ryan Nefdt, 'Conceptual Engineering: A Road Map to Practice', *Philosophy Compass* 17, no. 10 (2022): 1–15, 2, https://doi.org/https://doi.org/10.1111/phc3.12879.

[19] Common Heritage was the guiding concept for the deep seabed regime, under the United Nations Convention on the Law of the Sea (1982), and it was also mentioned explicitly in the Agreement Governing the Activities of States on the Moon and Other Celestial Bodies, better known as the Moon Treaty (1984).

[20] Surabhi Ranganathan, 'Global Commons', *European Journal of International Law* 27, no. 3 (2016): 693–717, 704, https://doi.org/10.1093/ejil/chw037.

economic returns, rather than legal propriety or ethical considerations for future generations.[21]

Just like with Permanent Sovereignty, nothing in principle prevents one from interpreting Common Heritage as emphasizing preservation over exploitation. In fact, one could argue that Common Heritage has lately been interpreted in precisely this way, for example, through UNESCO's list of World Heritage Sites. Historically, however, the latter sense has predominated.

In the middle of these two regimes stood Antarctica. To understand its uniqueness, it is useful to analyse what resource rights comprise. Chris Armstrong proposes one such typology (based in turn on Elinor Ostrom's work). When thinking about rights over resources, Armstrong claims, one should distinguish between two different levels: first-order rights or rights to ownership, and second-order rights or rights to jurisdiction, which shape and delimit the former. Rights to ownership over resources are the right to access, withdraw, alienate, and derive income from the resource in question. Rights to jurisdiction over resources are the right to exclude, manage, regulate alienation, and regulate income from the resource in question.[22] By virtue of Permanent Sovereignty, states enjoy full rights to jurisdiction over resources, but also rights to ownership over them, the limits of which depend on the way in which states themselves decide to draw the line. By virtue of Common Heritage, some suprastate body designated by states enjoys rights to jurisdiction and ensures that first-order rights are granted in such a way that, when it comes to the derivation of income, everyone benefits from the exploitation of resources (this is the role, for example, of the International Seabed Authority).

In the Antarctic case, the rights landscape looks different. States that claimed territory in Antarctica aspired to the first- and second-order rights characteristic of Permanent Sovereignty. The hoarding and controlling of resources for no immediate use and with no immediate purpose, as suggested in Chapter 3, was one of the wrongs of colonialism without people. Because the claims were never recognized, however, the claimants reached a compromise and agreed to change their normative situation. They froze the possibility of ruling individually over their Antarctic slices in exchange for jointly ruling over the whole continent. Consequently, they also had to freeze the possibility of claiming ownership-type rights to withdraw, alienate, and derive income from Antarctic resources.

[21] Joyner, *Governing*, 33–34.
[22] Armstrong, 'Against "Permanent Sovereignty"', 132–33.

Not surprisingly, the Mineral Convention took a decade of discussions because the Antarctic Treaty members had to delimit with precision who would get rights to withdraw the minerals (exclusion); how this would be done (management); on what basis rights to resources would be granted (regulation of alienation); and whether they would be levied (regulation of income).[23] If it had succeeded, it would have become a peculiar arrangement, granting a select group of states the right to manage the resources of a whole continent. It was doomed to failure by two groups of forces pulling in opposite directions. On the one hand, for claimant states, the Mineral Convention threatened their frozen, but not forgotten, aspiration to Permanent Sovereignty over those resources. On the other hand, for non-claimants and 'outsiders', the Mineral Convention threatened the possibility of turning Antarctica into a Common Heritage territory from whose exploitation all humans could benefit.

What happened instead was that, after the signature of the Environmental Protocol, a focus on territorial duties gained force. The Environmental Protocol banned the prospection and exploitation of mineral resources on the continent for 50 years subject to renewal, and human activity became highly regulated to minimize environmental impact. Member states retained their role as joint guardians over the whole continent, but with the important caveat that their participation was as guardians of peace and of the environment and generators of useful knowledge, rather than as potential beneficiaries of economic gains to be made from Antarctic exploitation. By banning the possibility of extracting resources, the rights to regulate alienation and income became irrelevant.[24] Moreover, the possibility that states or other entities claimed ownership-type rights to withdraw, alienate, and derive income from Antarctic resources remained suspended.

The six annexes to the Protocol (of which the first five are in force) reveal its ambitious scope: Environmental Impact Assessment, Conservation of Flora and Fauna, Waste Disposal and Management, Prevention of Marine Pollution, Area Protection and Management (including of historic sites and monuments), and Liabilities Arising from Environmental Emergencies.[25] There was one further element of the Protocol that buttressed the role of the parties as

[23] The process was somewhat analogous to the deep seabed negotiations leading to the United Nations Convention on the Law of the Sea, although here there were 14 parties, as opposed to all the UN member states.

[24] It must be remembered that this is the case for mineral resources. One could argue, however, that when it comes to activities like tourism, as already mentioned, there is a much greater need for regulation and arguably a need to tax it. It is surprising, to say the least, that there is no entry fee to visit the largest remaining wilderness in the world.

[25] Of these, the first five have been ratified. See Antarctic Treaty System, 'Protocol on Environmental Protection (1991)'.

joint guardians of Antarctica. This was the establishment of the Committee for Environmental Protection, already mentioned. The Committee is composed of representatives of the member states, accompanied by experts and advisors, and open to representatives of member states of the treaty but not the Protocol. It is also composed of the heads of the Scientific Committee on Antarctic Research (SCAR) and of the Scientific Committee for the Conservation of Antarctic Marine Living Resources, and other 'relevant scientific, environmental and technical organizations'.[26] Its role is to advise the Treaty members on a range of things including the effectiveness of measures and the need to update, strengthen, or improve them; 'the state of the Antarctic environment'; and 'the need for scientific research, including environmental monitoring'.[27]

In sum, one could say that the states that signed the Protocol contributed to the creation of a model of territorial governance where preservation was put front and centre, with human activities subordinated to it, rather than the other way around. I now turn to the prospects of using this model to think about territorial governance beyond Antarctica—specifically, in ecoregions around the world.

7.2 Joint Guardianship for ecoregions

In a provocative article, E. Dinerstein and colleagues propose the creation of a 'Global Safety Net' of 'ecoregions' which, if conserved, would contribute both to halting the biodiversity crisis and to mitigating climate change.[28] These ecoregions, which comprise half of the earth's terrestrial areas, are selected according to three criteria. The first is great diversity and abundance, measured by species rarity, distinct species assemblages, rare phenomena, and intactness (examples are the Madagascar humid forests, Borneo lowland rainforests, and the Scandinavian and Russian taiga). The second regards the potential for carbon storage and drawdown, in sites like the Kazakh steppe (Kazakhstan and Russia) and the New England Acadian forests (United States and Canada). The third refers to wildlife and climate corridors, like those currently in existence in the Northern tundra and taiga, and others in need of creation, to connect tropical forests with temperate and tropical grasslands. In terms of geographic distribution, the states with the greatest potential to contribute to the preservation

[26] Ibid., Article 11.
[27] Ibid., Article 12.
[28] Dinerstein et al., 'Global Safety Net'.

of ecoregions are Russia, Indonesia, Brazil, Canada, Australia, and the United States. In terms of population, it has been estimated that ecoregions are home to one billion people, with Indigenous communities occupying at least a third of their total area.[29]

There are three reasons why the preservation of ecoregions is a goal worth pursuing. First, as Dinerstein and colleagues mention, protecting these areas would help to achieve two objectives in one go: halting biodiversity loss and reducing CO_2 emissions globally. This can be justified even from a purely anthropocentric perspective, as self-interest in our own continuation. Being a species that has evolved amongst others for thousands of years, it is a big gamble for humans to simply assume that life will continue to be liveable for us if we get rid of (most of) the others.[30] Some might reply that humans have always managed to adapt to new circumstances, and that the situation today will be no exception to our successful adaptability; therefore, we need not worry so much about further losses. However, never in human history has the rate of exploitation and extinction been as high as now; what humans are causing today is not just a quantitative change (in the number of existing species), but a qualitative one, where the very functioning of basic systems that sustain our life on earth is put at risk. In the face of such risk, the rational thing to do is to suspend business as usual and take action.

Second, preserving ecoregions is preserving our potential for knowledge expansion. 'All men by nature desire to know', claimed Aristotle in the very first line of the *Metaphysics*.[31] And yet, today we are destroying at the highest speed in our history one of the main objects of that knowledge; namely, the planet's biodiversity. It has been estimated that, on top of the two million known species on earth, six million might still be unknown.[32] The wealth of information contained in them and in their interactions might be lost forever if we keep perpetuating the practices that are jointly responsible for what has been labelled 'the sixth extinction': habitat destruction (including that caused by climate change), pollution, overhunting and overfishing, introduction of invasive species, and population growth.[33] For the sake of knowledge, then, it seems imperative to preserve what took more than 4 billion years to evolve. It is

[29] See, respectively, Judith Schleicher, Julie G. Zaehringer, Constance Fastré, Bhaskar Vira, Piero Visconti, and Chris Sandbrook, 'Protecting Half of the Planet Could Directly Affect over One Billion People', *Nature Sustainability* 2, no. 12 (2019): 1094–96, and Dinerstein et al., 'Global Safety Net', 7.

[30] Edward O. Wilson, *Half-Earth: Our Planet's Fight for Life* (W. W. Norton & Company, 2016).

[31] Aristotle, 'Metaphysics', in *The Complete Works of Aristotle, Bollingen Series LXXI-2*, ed. Jonathan Barnes (Princeton University Press, 1984), 1552.

[32] Wilson, *Half-Earth*, 21.

[33] Ibid., 57–58.

the very material of scientific knowledge and human curiosity and inquisitiveness that is at stake.

Third, and more importantly, for those who are dissatisfied with the implications of methodological anthropocentrism for our treatment of nature, it is obvious that preserving ecoregions is something we owe not just to present and future humans, but also to the ecoregions themselves and their non-human inhabitants. I have suggested in Chapters 5 and 6 why I think that a normative framework where everything that is not human is purely a means for us is inadequate, and why there are good reasons to incorporate nature as part of the political community.

Currently, all terrestrial ecoregions in the world fall under the regime of Permanent Sovereignty. This is inadequate from an environmental point of view because it turns them into fragmented territories under the governance of different states, which often fail to coordinate among themselves and may even take measures that are to the direct detriment of ecoregions. Furthermore, as mentioned above, under Permanent Sovereignty, exploitation—rather than preservation—is usually understood as the default. To take one obvious example: in May 2020, the Brazilian president, Jair Bolsonaro, transferred the regulation of public forests from the Ministry of the Environment to the Ministry of Agriculture, opening the forest to 'development' (i.e. industrial cattle-raising and monoculture plantations). When consulted about the potential exploitation of the Amazon and its use for mining and development, Bolsonaro invoked Permanent Sovereignty to justify his decision: 'We understand the importance of the Amazon for the world—but the Amazon is *ours* . . . We preserve more [rainforest] than anyone. No country in the world has the moral right to talk about the Amazon. You [Europeans] destroyed *your own* ecosystems'.[34] The tacit and worrying conclusion of Bolsonaro's *tu quoque* argument is that Brazil may destroy its 'own' ecosystems now (just like Europeans destroyed 'theirs' before). These declarations happened at a time when forest fires were raging in the Amazon, with scientists pointing to deforestation as one of the main causes.[35] Furthermore, although the Amazon forest extends over eight different states and one overseas territory, the negative effects of the decisions

[34] Dom Phillips, 'Bolsonaro Declares "The Amazon Is Ours" and Calls Deforestation Data "Lies"', *The Guardian Online*, 19 July 2019, my emphases, https://www.theguardian.com/world/2019/jul/19/jair-bolsonaro-brazil-amazon-rainforest-deforestation.

[35] Vânia R. Pivello, Ima Vieira, Alexander V. Christianini et al., 'Understanding Brazil's Catastrophic Fires: Causes, Consequences and Policy Needed to Prevent Future Tragedies', *Perspectives in Ecology and Conservation* 19, no. 3 (2021): 233–55, https://doi.org/10.1016/j.pecon.2021.06.005.

made for the 'Brazilian' Amazon also affected these neighbouring countries and—one should add—the whole world.

With this in mind, I suggest looking at the Antarctic arrangement not as a source of direct emulation, but as a source of inspiration for novel experiments in Joint Guardianship.[36] Ecoregions, like Antarctica, are vital for the functioning of earth systems that keep the planet inhabitable for humans and for a myriad of other species. Their preservation, like that of Antarctica, requires an integrated approach that does not follow the rigid lines of political maps, but much more diffuse ecological borders. So, what elements may be suited for adaptation here?

When it comes to the Antarctic Treaty, the cornerstone of the whole edifice is—as has been repeatedly mentioned—the suspension of actual and potential territorial claims by states. The normative situation for individual states acting as joint guardians over ecoregions, analogously, would be the following. Instead of holding first- and second-order rights over ecoregions that happen to fall within their territories (as is the case today), they would put their ownership-type rights to the side insofar as this is needed for the successful protection of the place. As for their jurisdiction-type rights, they would share them with other states, and also ideally with other actors, thus moving beyond the assumptions of methodological nationalism (more on this in the following section).

The Environmental Protocol designates Antarctica as 'a natural reserve, devoted to peace and science',[37] and declares the protection of the Antarctic environment as a fundamental consideration when deciding what kinds of activities may be undertaken there, and in what manner. Ecoregions should be devoted to similar ends, and their access and use should be subordinated to these ends. A Committee for Environmental Protection akin to that of the Antarctic Treaty could be created to give advice on how best to achieve these ends, while the establishment of periodic inspections by observers would ensure that the agreed-upon regulations are complied with. Eventually, this Committee could acquire decision-making power.

A key factor in explaining the role of humans in climate change and biodiversity loss is what Simon Caney calls 'harmful short-termism': presentist thinking and policy making that 'results in governments violating their

[36] As Finn Sollie once expressed: 'while the complete structure of Antarctic cooperation may not lend itself to transfer to other regions, elements of it may be suited for adaptation to new environments': Finn Sollie, 'The Political Experiment in Antarctica', in *Frozen Future*, ed. Richard S. Lewis and Philip M. Smith (Quadrangle Books, 1973), 49.

[37] Antarctic Treaty System, 'Protocol on Environmental Protection (1991)', Article 2.

responsibilities to current and future generations to create just and ecologically sustainable societies'.[38] Agreeing on Joint Guardianship for ecoregions (and maybe, eventually, on a Global Environmental Protocol for Ecoregions) would represent a move away from that kind of thinking and towards an institutional design that is viable not just for those of us living today, but also for our descendants and for nature as such.

7.3 Four objections

In what follows, I address four potential objections to the idea of implementing Joint Guardianship over ecoregions. They all point to substantive disanalogies between ecoregions and Antarctica. I suggest that, while they raise genuine concerns, these objections point more to the fact that Joint Guardianship should be sensitive to specific contexts, rather than to the impossibility of applying the model beyond Antarctica.

7.3.1 Unfair to states

First, because the territorial claims of individual states in Antarctica were never recognized, what these states gave up by signing the Treaty was not actual sovereignty, but the dream of achieving it. In other words, in Antarctica sovereign claims were frozen before they were effectively realized. Under the scheme suggested here, on the contrary, states would have to give up claims over areas that have historically been recognized as theirs. My proposal would thus be problematic in that it takes away rights already in place.

My response to this is that what might have been a just arrangement at some point in the past may stop being so in the present, due to deeply changed circumstances.[39] Among these are a clearer understanding of how basic earth systems work and how distant ecosystems influence each other; global population growth; and a rocketing use of nature for human purposes. If seventy years ago Permanent Sovereignty was thought to be the best path to regulate

[38] Simon Caney, 'Democratic Reform, Intergenerational Justice and the Challenges of the Long-Term', Centre for the Understanding of Sustainable Prosperity, 2019, 3, https://www.cusp.ac.uk/wp-content/uploads/11-Simon-Caney-online.pdf.

[39] This is the inverse of 'supersession of injustice', where what might have been unjust in the past becomes just in the present given profoundly changed circumstances. See Jeremy Waldron, 'Superseding Historic Injustice', *Ethics* 103, no. 1 (1992): 4–28.

affairs between states in a peaceful manner, as well as enabling each of them to improve the well-being of their human inhabitants, it is high time to reconsider this function. International law is a dynamic field, and it will continue to evolve as we get a better grasp of the fact that the political division of the world into discrete territorial units with extensive sovereign rights is environmentally (and normatively) inadequate. There are a growing number of voices pushing for this change. In the same spirit of the proposal for a 'Global Safety Net' for ecoregions, the proponents of the concept of planetary boundaries now advocate for a system of governance that puts the 'planetary commons' (of which Antarctica is one) centre stage.[40]

Still, some may worry that this proposal is unfair in a way that the Antarctic Treaty was not. Because of the disputed status of territorial claims in Antarctica, the idea of prohibiting anyone from benefiting economically from its resources was, arguably, the fairest attainable outcome. Here, on the contrary, the rules of the game would be changed so that what used to be a legitimate source of income for states would no longer be so. This would be especially objectionable for developing countries that rely on resource extraction to lift their populations above the poverty threshold.

I think that this objection only holds if one grants two claims that are not self-evident: that the current model of economic growth is sustainable, and that states as they are drawn on the map today are legitimate owners and sovereigns over their territories—in other words, that the current division is morally adequate.

On the first point, demands for alternative economic models that establish a different relationship with the environment have been as strong in developing countries as in developed ones.[41] This suggests that the picture of neocolonial powers imposing their conservation obsessions upon their neocolonies—by impeding the development path they themselves pursued—is deceiving and overly simplistic. Furthermore, under Joint Guardianship administrative responsibilities and economic costs would be shared. To the extent that we acknowledge that individual states are not exclusive owners of their land and resources, we should also acknowledge that the responsibilities of preserving

[40] Johan Rockström, Louis Kotzé, Svetlana Milutinović et al., 'The Planetary Commons: A New Paradigm for Safeguarding Earth-Regulating Systems in the Anthropocene', *Proceedings of the National Academy of Sciences* 121, no. 5 (2024): e2301531121, https://doi.org/10.1073/pnas.2301531121.

[41] For alternative models, see J. Blewett, and R. Cunningham, eds., *The Post-Growth Project: How the End of Economic Growth Could Bring a Fairer and Happier Society* (Green House, 2014); Giorgos Kallis, Vasilis Kostakis, Steffen Lange, Barbara Muraca, Susan Paulson, and Matthias Schmelzer, 'Research on Degrowth', *Annual Review of Environment and Resources* 43, no. 1 (2018): 291–316; and Arturo Escobar, 'Degrowth, Postdevelopment, and Transitions: A Preliminary Conversation', *Sustainability Science* 10, no. 3 (2015): 451–62, https://doi.org/10.1007/s11625-015-0297-5.

them do not fall exclusively upon individual states or local communities. It would be unfair, indeed, to charge biodiversity-rich states like Brazil and Indonesia with the full economic, administrative, social, and political bill for a job that benefits everyone and where everyone has a stake. Theorists writing on conservation ethics have proposed basic principles for how to share the costs in a fair way, and I will not dwell on them here, except by noting that natural capital has to be given its full value in these calculations.[42] As noted by the economist Partha Dasgupta, for centuries we have used and abused the various goods and services provided by nature at no monetary cost to ourselves. This is not just a market failure, but also a political failure that needs to be corrected by a deep reform of the way our institutions currently work. Today, governments give incentives to exploit rather than preserve, they subsidize environmentally damaging industries, and they fail overall to protect global public goods.[43] Moving forward, governments should give incentives to protect rather than to exploit, they should cut subsidies for environmentally damaging industries, and they should aim for the protection of global public goods, of which ecoregions are a prime example. If conservation is still considered a 'burden', in short, it is only because unfettered development makes it seem like one. What should be questioned are the purported virtues of development rather than the potential challenges that conservation poses.

On the second point, namely, that states as they exist today are legitimate sovereigns over their territories, it is useful to remind readers why this is not self-evident. As mentioned in earlier chapters, arguments used to justify territorial rights come in two main kinds: on the one hand, connection-based arguments that appeal to some normatively relevant link between the territorial agent and the space claimed; and, on the other hand, function-based arguments that invoke some key functions performed by the territorial agent that

[42] Chris Armstrong proposes three principles based on *contribution* to the problem (namely, to the destruction of biodiversity); *benefits* derived from preserving the place; and *ability to pay*. Chris Armstrong, 'Sharing Conservation Burdens Fairly', *Conservation Biology* 33, no. 3 (2019): 554–60, 554, http://dx.doi.org/10.1111/cobi.13260. On natural capital and how to measure its value, see Edward Barbier, 'The Concept of Natural Capital', *Oxford Review of Economic Policy* 35, no. 1 (2019): 14–36 and Robert Costanza, Ralph D'Arge, Rudolf De Groot et al., 'The Value of the World's Ecosystem Services and Natural Capital', *Nature* 387, no. 6630 (1997): 253–60, https://www.nature.com/articles/387253a0.pdf. Here some might object that talking about *natural capital* commodifies nature and makes us fall back on the same way of thinking that we are trying to move away from. I think this is right if we see nature merely as a commodity, but not if we acknowledge that it has value beyond that, such that we need to account for its responsible use.

[43] Partha Dasgupta, 'The Economics of Biodiversity: The Dasgupta Review—Headline Messages', UK Government Publishing Service, February 2021, 2, https://assets.publishing.service.gov.uk/media/60182857d3bf7f70c2afe5bb/Dasgupta_Review_-_Headline_Messages.pdf. For an extensive analysis of these issues, see Partha Dasgupta, *The Economics of Biodiversity* (Cambridge University Press, 2024), https://doi-org.ezproxy.uio.no/10.1017/9781009494359.

require the latter to have territorial rights. In the case of Antarctica, I showed why not even in the most generous reading of these arguments were claimants justified in extending their claims in the way they did. The same holds, I suggest, for many claimed areas beyond Antarctica. Although I cannot conduct a detailed analysis showing why this is the case here, I hope that the following example will suffice to make the point.

Because they depend upon an actual, morally relevant connection between the territorial agent and the place, the plausibility of connection-based theories is directly proportional to the distance between the people and the *homeland* or *heartland*.[44] This means that, while connection-based arguments sound commonsensical when justifying, for instance, why Canada has territorial rights over Ottawa, their plausibility dilutes the further away one moves from the populated areas—that is, they have trouble justifying why the furthest reaches of the Northwest Territories should also be under the full sovereignty of the Canadian state. Acknowledging this, some theorists have suggested that 'peripheral' land and 'peripheral' natural resources should not be treated as being owned by states, but rather as being under the state's care. This is Margaret Moore's proposal for Canada in the Arctic: while the traditional arguments for full Canadian sovereignty over most Arctic land and resources are found wanting, Moore suggests that there might still be good grounds to keep the region under Canadian stewardship.[45] While I agree with Moore that stewardship (or guardianship) rather than sovereignty should be the key word for understanding the relationship between Canada and its furthermost reaches, I disagree that it should be Canada alone that fulfils that role, especially when the decisions made affect those beyond Canadian borders.

When it comes to function-based arguments, their plausibility depends on how direct the relationship between the function fulfilled and the need for territorial rights is. If the aims are to establish justice and to guarantee security and the fulfilment of (at least) the basic needs of the population, the territorial agent needs control over the territory to be successful. Once justice is established and basic needs are satisfied, however, it is hard to keep justifying the extension of territorial rights based on function.[46] What function-based arguments thus

[44] See, respectively, Miller, 'Territorial Rights,' 261; Moore, *Political Theory of Territory*, 118–22.

[45] The stewardship justification, Moore clarifies, 'does not support the usual understanding of resource rights, which permits resource use and resource-extraction activities, though stewardship clearly does involve control over natural resources': Margaret Moore, 'Is Canada Entitled to the Arctic?' *Canadian Journal of Philosophy* 50, no. 1 (2020): 98–113, 99, https://doi.org/DOI:%252010.1017/can.2019.8. Moore thus thinks that the Canadian state should continue to hold exclusive jurisdiction, but not necessarily ownership rights, over these areas.

[46] Nine, *Global Justice*, 42.

share with connection-based ones is that they are ill-equipped to justify states' Permanent Sovereignty over resource-rich areas, when the population of those states is relatively small and not even geographically close to those resources. Staying with the example of Canada, it is hard to make a plausible function-based argument that the state needs over a tenth of the world's total landmass to guarantee the establishment of justice, the provision of security, and the satisfaction of basic needs for 0.48 percent of the total world population, mostly located in the vicinity of the border with the United States.

To be sure, some may point out that states have historically sought to extend their *Lebensraum* and that, as a result, most current state borders are to a lesser or greater extent arbitrary. Yet, all in all, it is better for world peace to have a global political order where all land is under some state's jurisdiction, where geographical borders are no longer shifting, and where territorial conquest is no longer permissible.[47] Furthermore, given recent developments in our world today (e.g. Russia's invasion of Ukraine), one might think that holding to this status quo is absolutely essential. To this I would respond that I am not suggesting that states with territories that are unjustified by any plausible connection or function-based argument should return them to the 'commons', or that we should try to fix past injustice through territorial redistribution. What I am suggesting is that the impossibility of justifying the full extension of their territories should make states open and flexible to the possibility of sharing some rights over them (specifically, second-order rights) in conjunction with others, when the goal is the protection of ecosystems of global importance.

7.3.2 Unfair to people

A second, obvious difference between Antarctica and ecoregions for the purposes of applying Joint Guardianship is that while there are no human Antarcticans, there are one billion people living in the ecoregions where the Permanent Sovereignty of individual states would be frozen (around 25 million in the Brazilian Amazon alone). For these people, losing rights over the place they inhabit might be seen as a new colonial wave.[48]

[47] A persuasive defence of the territorial status quo is found in Leif Wenar, *Blood Oil: Tyrants, Violence and the Rules That Run the World* (Oxford University Press, 2016).

[48] See, for example, Jenni Monet, '" Green Colonialism": Indigenous World Leaders Warn over West's Climate Strategy', *The Guardian Online*, 23 April 2023, https://www.theguardian.com/world/2023/apr/23/un-indigenous-peoples-forum-climate-strategy-warning.

But Joint Guardianship should not be confused with 'fortress conservation', which sees humans in necessary opposition to nature and thus seeks to remove all human presence from areas designated as 'wilderness'.[49] In fact, one aspect where Joint Guardianship over ecoregions could improve over the Antarctic model is precisely this: by incorporating as guardians not just other states, but also supranational governmental and non-governmental entities, local communities, and (as suggested in Chapters 5 and 6) representatives of non-humans and of nature. The information required to successfully manage complex ecosystems is distributed among many different actors. This means that effective institutional structures must 'pool knowledge and perspectives among and across different levels—global, regional, national and local—and from different organizations, communities and individuals'.[50] Under Joint Guardianship, accordingly, the place of local communities and, among these, of Indigenous communities, would be underlined, especially since there is a considerable overlap between ecoregions and areas inhabited by Indigenous groups.[51] As opposed to what happens today, where Indigenous groups are at best consulted and at worst ignored when it comes to making crucial decisions about the future of these areas, under Joint Guardianship they would have a much more prominent jurisdictional role. Against the worry that Joint Guardianship would become a new version of 'fortress conservation', thus, local communities would retain their right to inhabit ecoregions and use them for their living as most of them have been doing for centuries, in nonabusive ways that allow for renewal.[52]

Another reason why non-state actors should embrace Joint Guardianship is that while the idea of Permanent Sovereignty was demanded and celebrated above all by the central elites of newly independent states, the relationship between them and Indigenous groups and groups in the 'peripheries' (where most ecoregions lie) has never been an easy one. That many peoples within states do not feel part of them and oppose the imposition of legal structures that seem

[49] On fortress conservation as neocolonialism, see William M. Adams, 'Nature and the Colonial Mind', in *Decolonizing Nature: Strategies for Conservation in a Post-Colonial Era*, ed. William M. Adams and Martin Mulligan (Earthscan, 2003), https://doi.org/10.4324/9781849770927. For examples of injustice perpetrated in the name of conservation, see Duffy, *Nature Crime.*

[50] Dasgupta, 'Economics of Biodiversity', 4.

[51] Claudia Sobrevila claims that the ancestral lands of Indigenous peoples contain 80 per cent of the remaining healthy ecosystems and global diversity priority areas, including rainforests: Sobrevila, *Role of Indigenous Peoples*, 5.

[52] As for inhabitants of ecoregions who reject Joint Guardianship and invoke their right to 'develop', one might question the legitimacy of such a right. Just as states may expropriate individual owners for the common good, so should guardianship arrangements be able to restrict the uses of land by local inhabitants and offer the possibility of relocation or some form of compensation for those who are unwilling to protect and are not able to show that exploitation is justified.

alien to them strengthens rather than weakens the Joint Guardianship proposal. Protecting the planet is often presented as something that will affect people largely negatively, but this is only because it is assumed that environmental protection will be done through the usual channels, without the participation of those who have long been owed a more active role. Furthermore, the concept of guardianship is much more familiar to many of these groups than the concept of sovereignty, and much more aligned with the values that some Indigenous scholars have claimed are the central values of many Indigenous ethics.[53]

7.3.3 Unfeasible

A third potential objection against Joint Guardianship is that it seemingly ignores what political theorists call 'feasibility constraints'; namely, considerations of how realizable the execution of an idea should be for it to be taken seriously.[54] So, does it make sense to propose a model that the main actors presumably will not be motivated to follow? As already noted in Chapter 6, I think the answer to this question depends on how feasibility constraints are defined. Following Pablo Gilabert, I propose to understand 'soft' feasibility constraints (those that have to do with psychological, social, political, and economic causes, rather than laws of logic or physics, which are 'hard') as malleable through our political imagination and our commitment to change current circumstances. Principles that seem impossible to fulfil today may thus become possible to achieve in the future—through the performance of what Gilabert calls 'dynamic duties'.[55] On this reading of feasibility, from the fact that states are not motivated today to give up part of their current package of rights it does not follow that they will not be willing to give them up at a later point in time. Rather, if there are strong, substantiated reasons that show that the current arrangement is untenable, then it is up to the different political agents—from individual citizens to non-governmental organizations to state and intergovernmental institutions—to take a transitional standpoint and strive to move in the direction of an alternative, more desirable arrangement.[56] This is already happening both among citizens of states where ecoregions lie and of states which continue

[53] See, for example, John Borrows, *Law's Indigenous Ethics* (University of Toronto Press, 2019).
[54] The literature on feasibility has burgeoned in the last decade. For an overview, see Southwood, 'Feasibility Issue'.
[55] Gilabert, 'Justice and Feasibility', 95.
[56] Ibid., 113ff. See also what Elizabeth Cripps calls 'promotional duties', that is, those 'to attempt to bring about the necessary collective action': Elizabeth Cripps, *Climate Change and the Moral Agent: Individual Duties in an Interdependent World* (Oxford University Press, 2013), 140.

to profit from oil extraction.[57] The mere proposal of an alternative model like Joint Guardianship might help to showcase the inadequacy of Permanent Sovereignty more clearly, and help the needed changes to gain momentum.

Furthermore, even hardcore realists must admit that the package of states' sovereign rights has been constantly shrinking during their relatively short historical lifetime. If territory, with people on it, could be sold a hundred years ago, this is not the case anymore, even though some would like to go back to that era. If states used to think of the atmosphere as an endless sink in which to dump their waste, a majority of them have committed themselves to restricting their use of it.[58] Aligned with the goal of protecting nature, moreover, in 2021 the G7 committed themselves to the protection and conservation of at least 30 percent of global land and 30 percent of global oceans by 2030. Leading by example, these states pledged to protect at least 30 percent of their own territories by that year.[59] Giving up ownership rights and being willing to share jurisdictional rights would still be a major step ahead. But here one should also bear in mind that social and political transitions are not linear, but subject to tipping points. This suggests that scenarios that may look impossible to achieve today might be just around the corner, depending on the changes we make now.[60]

7.3.4 Why not other models?

Finally, some might be wondering why I did not select other already-existing models of governance that might seem even better suited to these regions than

[57] See, for example, Coalizão Brazil, a coalition of over two hundred NGOs and agribusinesses fighting against the deforestation of the Brazilian Amazon (Brazilian Coalition on Climate, Forests and Agriculture, 'About Us', https://coalizaobr.com.br/en/about-us/) and Future in Our Hands, a Norwegian NGO that campaigns for putting a stop to oil extraction and demands a transition to renewable sources of energy (Future in Our Hands, 'Join the Movement for a Green Future', https://www.framtiden.no/en).

[58] United Nations, 'Paris Agreement', 2015, https://unfccc.int/sites/default/files/english_paris_agreement.pdf.

[59] Campaign for Nature, 'G7 Leaders Agree to Historic "Nature Compact"—Set Comprehensive Biodiversity Targets, Commit to Protecting At Least 30% of Lands and Seas', Media statement, 13 June 2021, https://www.campaignfornature.org/g7-nations-agree-to-historic-nature-compact. The United States stepped down from both the Paris Agreement and the 30×30 pledge in 2025, but some federal states have decided to keep their commitments despite the US decision. Will 30x30 succeed? I cannot deny that world politics today makes environmental cooperation and goodwill among states look endangered. At the same time, I think it is important to insist in these values precisely when they seem to be retreating—as confirmed by the reaction of states like California to keep their conservation goals, and the US Climate Alliance (a net-zero focused group of twenty four state governors) to stay committed to the goals of the Paris Agreement.

[60] Concrete proposals for how to activate these tipping points are found in Ilona M. Otto, Jonathan F. Donges, Roger Cremades et al., 'Social Tipping Dynamics for Stabilizing Earth's Climate by 2050', *Proceedings of the National Academy of Sciences* 117, no. 5 (2020): 2354–65, https://doi.org/10.1073/pnas.1900577117.

the Antarctic one. Among them: the above-mentioned UNESCO's World Heritage Program, the Svalbard Treaty, and public-private partnerships for conservation purposes. However, an obvious feature of these alternatives is that they all take Permanent Sovereignty for granted, and therefore the consequent allocation of ownership rights to individual states. Thus, they remain inadequate. Article VI of the World Heritage Convention (1972) clearly underlines that all measures will be taken 'fully respecting the sovereignty of the States on whose territory the cultural and natural heritage mentioned . . . is situated, and without prejudice to property right provided by national legislation'.[61] In the case of the Svalbard Treaty, the driving principle is equal access to, and ownership of, resources (including mineral rights); that is, the focus is on exploitation rather than preservation, however 'sustainable' that exploitation might be.[62] As for private enterprises administering ecoregions, if there is any fear of colonialism or neocolonialism re-emerging for the purposes of conservation, this is where it should arise: with states giving up their management rights to entities the main aim of whom is to make business profitable for their shareholders, and with local inhabitants becoming part of the landscape.[63]

7.4 Concluding remarks

In this chapter, I have suggested that the Antarctic territorial model may provide a blueprint for the governance of other regions of the world—specifically, but not exclusively, ecoregions. Subject to neither Permanent Sovereignty nor Common Heritage, Antarctica has remained the only continent on earth where business has not been done as usual, with states focusing on guardianship rather than exploitation, and with collaboration and coordination among them being the rule and not the exception. At the same time, looking at the Antarctic Treaty through the lenses of other places, like ecoregions, helps us see what has been missing and what may be improved. For one thing, an adequate model of territorial governance requires moving beyond the assumptions of methodological nationalism, something that is yet to happen in international law. Given

[61] UNESCO, 'Convention for the Protection of the World Cultural and Natural Heritage', 16 November 1972, https://unesdoc.unesco.org/ark:/48223/pf0000002091.

[62] 'Treaty Between Norway, the United States of America, Denmark, France, Italy, Japan, the Netherlands, Great Britain and Ireland and the British Overseas Dominions and Sweden Concerning Spitsbergen Signed in Paris 9th February 1920', Arctic Portal, http://library.arcticportal.org/1909/1/The_Svalbard_Treaty_9ssFy.pdf.

[63] On how this model of privately managed national parks has been growing in Africa, see '"Elephants" Graveyard No More', *The Economist*, 24 October 2020, 31–32.

the complexity of the issues at stake and the plurality of views around them, I have suggested that what is needed is multilevel engagement that includes not just states, but also supra- and international bodies, as well as non-state entities and local communities. To move beyond the framework of methodological anthropocentrism, moreover, these new models of territorial governance ought also to include representatives of non-humans and nature, especially when the decisions taken affect them in relevant ways. In sum, the aim should be to move towards a system of Joint Guardianship where not just states have decision-making power and where not only humans are represented.

8
Facing the Paradox of Protection

If the history of the Antarctic Treaty were to be marked by some major event, the Environmental Protocol negotiated in Viña del Mar in 1990, and signed in Madrid in 1991, would be the most suitable candidate. Before the Environmental Protocol, it was routine for the scientific stations to throw their garbage bags into the sea. Heavy machinery was left on ice to either sink or drift away. Walking through rare moss communities and disturbing penguin colonies were seen as normal conduct and not frowned upon. Small oil spills, loss of fuel gallons, and careless disposal of human waste and gear happened without sanction, and dogs were common members of expeditions.[1] All this changed after the Environmental Protocol, when a 'leave no trace' ethos became the norm and when environmental regulations became increasingly strict and increasingly 'accepted and expected'.[2] Since then, the 'ten-metre rule' between humans and Antarctic fauna became the norm, unless scientists need to conduct research directly on the animals or unless the animals themselves decide to get friendly. Specific rules were also developed for specific sites. In the Specially Managed Area of the McMurdo Dry Valleys, for example, there is an official rule that scientists working there must restore any rocks they move to their original location, to preserve the apparent intactness of the place and leave nature 'as they found it'.[3]

The comprehensiveness and specificity of the Protocol's regulations have turned environmental protection into the fourth 'considerable value' of Antarctica—together with the suspension of the territorial claims, the preservation of peace, and scientific cooperation.[4] At the same time, the rules to minimize human impact *in* Antarctica are clearly insufficient for the protection *of* Antarctica—a phenomenon that is made obvious by climate change. Together with Greenland, Antarctica is the most affected place on earth because of the

[1] Jessica O'Reilly, *The Technocratic Antarctic: An Ethnography of Scientific Expertise and Environmental Governance* (Cornell University Press, 2017), 38. See also Tina Tin, Z. L. Fleming, K. A. Hughes et al., 'Impacts of Local Human Activities on the Antarctic Environment', *Antarctic Science* 21, no. 1 (2009): 3–33, http://journals.cambridge.org/abstract_S0954102009001722. Dogs were banned from Antarctica to prevent the spread of diseases and to protect native wildlife.

[2] O'Reilly, *Technocratic Antarctic*, 39.

[3] Ibid., 53.

[4] Hemmings, 'Considerable Values'.

From Sovereignty to Guardianship. Alejandra Mancilla, Oxford University Press. © Alejandra Mancilla 2026.
DOI: 10.1093/9780197902769.003.0008

'polar amplification' of climate change. Large parts of West Antarctica, which is mostly composed of ice shelves floating on the sea, are doomed to melt[5]; penguin species like Emperors are in serious trouble due to record-low sea ice levels,[6] and non-native species pop up in the growing ice-free areas.[7] More than any other place on earth, Antarctica reveals 'the paradox of protection': to protect a territory, it is not enough to act only *within* that territory; action *beyond it* is just as—or even more—important.[8] Furthermore, as Peder Roberts points out, because of climate change Antarctica becomes not only a place that we want to protect because of its unique beauty and scientific value, but also because it itself becomes a threat: sea-level rise in the coming decades will be mostly the result of Antarctica melting, and will force the displacement of millions of people living in low-lying coastal areas around the planet.[9]

In this chapter, I show that the Environmental Protocol was drafted and signed at a time when the palpable impact of humans on the earth's atmosphere and environment was already clear, and when the rising levels of atmospheric greenhouse gases, the accelerated rate of species extinction, and the overuse of natural resources globally were matters of growing attention. During the preparatory meetings and discussions, the goal of keeping Antarctica as an unspoiled wilderness for conducting science useful for all was mentioned as much as the goal of understanding and keeping Antarctica as a key regulatory force within our earth systems. Starting from these documents, I suggest that a more consistent interpretation of the Environmental Protocol (together with some of its amendments) could ameliorate its current problem of fit, which arises because of the incongruity between the way in which the ecosystem functions and the characteristics of the institution designed to protect it.[10] This

[5] Alastair G. C. Graham, Anna Wåhlin, Kelly A. Hogan et al., 'Rapid Retreat of Thwaites Glacier in the Pre-Satellite Era', *Nature Geoscience* 15, no. 9 (2022): 706–13, https://doi.org/10.1038/s41561-022-01019-9 and Ted A. Scambos, Robin E. Bell, Richard B. Alley et al., 'How Much, How Fast?: A Science Review and Outlook for Research on the Instability of Antarctica's Thwaites Glacier in the 21st Century', *Global and Planetary Change* 153 (2017): 16–34.

[6] Philip N. Trathan, Barbara Wienecke, Christophe Barbraud et al., 'The Emperor Penguin: Vulnerable to Projected Rates of Warming and Sea Ice Loss', *Biological Conservation* 241 (2020): 108216, https://doi.org/10.1016/j.biocon.2019.108216 and Stéphanie Jenouvrier, Marika Holland, Julienne Stroeve et al., 'Projected Continent-Wide Declines of the Emperor Penguin Under Climate Change', *Nature Climate Change* 4, no. 8 (2014): 715–18, https://doi.org/10.1038/nclimate2280.

[7] Peter Convey and Lloyd S. Peck, 'Antarctic Environmental Change and Biological Responses', *Science Advances* 5, no. 11 (2019): eaaz0888, https://doi.org/10.1126/sciadv.aaz0888.

[8] Alejandra Mancilla and Peder Roberts, 'The Antarctic Paradox', Aeon, 27 May 2022, https://aeon.co/essays/how-do-we-solve-the-paradox-of-protection-in-antarctica.

[9] Peder Roberts, 'Does the Science Criterion Rest on Thin Ice?', *Geographical Journal* 189 (2023): 18–24, https://doi.org/10.1111/geoj.12367.

[10] Oran R. Young, 'Institutions and Environmental Change: The Scientific Legacy of a Decade of IDGEC Research', in *Institutions and Environmental Change*, ed. Oran R. Young, Leslie A. King, and Heike Schroeder (MIT Press, 2008).

would allow it to fulfil its goal of 'comprehensive protection of the Antarctic environment and dependent and associated ecosystems'.[11] In this interpretation, the parties would have to commit themselves not just to changing their conduct south of 60 degrees south but also at home. I suggest that the Antarctic Treaty members have a privileged opportunity to surmount the paradox of protection, because it is in their power to protect the continent by acting *in* it and *beyond* it. This possibility, on the contrary, is currently precluded in other places affected by this paradox, like Small Island States, which need the collaboration of others to successfully protect their own territories, but where this collaboration is usually seen as optional rather than as obligatory. I address a potential objection; namely, that environments are constantly changing, and that profound changes in Antarctica might be inevitable, even if we put all our efforts into protecting it. I conclude that, while the Environmental Protocol has been relatively successful at protection *in situ*, this must be complemented with *ex situ* action. We might not be able to guarantee that there will not be disruptive changes, but we can ensure that we do our best given the best available knowledge.

8.1 The object and purpose of the Environmental Protocol

As mentioned in previous chapters, the 1980s was the decade during which the Antarctic Treaty Parties negotiated a Mineral Convention that never came to be. It was also the decade when the Question of Antarctica was raised at the United Nations General Assembly, when states that were not members of the Antarctic Treaty criticized its club-like features and demanded a fair share of the economic benefits if the continent was to be turned into a site for resource exploitation. The 1980s was, finally, the decade when the most momentous decision for Antarctica was taken, by placing environmental protection front and centre, keeping a whole continent beyond the logic of resource extraction, with human activities tightly regulated and constrained. It is this display of self-restraint that makes the Antarctic case inspiring and worth imitating. For once, states came together and agreed that preservation rather than exploitation should be the goal; that nature had value for its own sake, and that the role of scientific experts in territorial governance was not optional, but essential.

[11] Antarctic Treaty System, 'Protocol on Environmental Protection (1991)', Article III.

To those prone to cynicism, the previous paragraph may sound like a piece commissioned by Antarctic Treaty members. But it is not. What interests me is what states committed themselves to doing, even though sometimes what is done is achieved more by omission than by action. To be sure, some could say that Antarctica was a low-hanging fruit for environmental protection; an easy case where states could show themselves to be environmentally friendly at small cost and presumably with considerable benefit. This agreement, they would add, could have never happened for the protection of other places; it did not even include the Southern Ocean, where fishing has been an ongoing activity. Yet, while all this might be true, states had a choice, and their choice was to exclude Antarctica from the sphere of business-as-usual. That fact is what I wish to underline here, and what I think should become a source of inspiration for territorial governance in places beyond Antarctica.

To understand the object and purpose of the Protocol, it is instructive to look at the genesis of its idea in a draft recommendation presented by Australia and France during the Antarctic Treaty Consultative Meeting of 1989. This proposal led to three special meetings over the next two years, during which the Protocol was drafted, and its title became the main goal of the Protocol: 'Comprehensive measures for the protection of the Antarctic environment and its dependent and associated ecosystems'.[12] This document mentions three distinctive values of Antarctica—values that were further reaffirmed by the heads of the delegations attending the special meetings. First, 'Antarctica has a major influence in the world climate . . . [and] a major influence in the global environment . . . Any significant changes to the Antarctic environment could have global consequences for atmospheric and oceanic circulation and upon sea levels'.[13] Second, 'the Antarctic environment offers special opportunities for research in many fields', especially for those aimed at understanding the global environment.[14] And third, 'Antarctica presents an unsurpassed opportunity for wilderness preservation . . . There is a value in the preservation of those qualities for their own sake'.[15] In other words, Antarctica's protection was important because of its role as a global thermostat and regulator of ocean and atmospheric circulation; its utility for science, especially science geared at understanding Antarctica's role in those earth systems; and its status as a unique wilderness—a status that might be worth preserving independently of any instrumental considerations.

[12] Australia and France, 'Joint Australia/French Proposal'.
[13] Ibid., 1.
[14] Ibid.
[15] Ibid., 2.

These three values were explicitly recognized in the Protocol. As has been discussed earlier, the intrinsic value of Antarctica was mentioned in Article 3, on environmental principles, as was its value as an open laboratory for science. Its value as a 'systemic commons' or 'planetary commons', meanwhile, was recognized indirectly in the preamble and Article 3.[16] There, it says that Antarctica offers unique opportunities 'for scientific monitoring of and research on processes of global as well as regional importance',[17] and research in Antarctica is mentioned as 'essential to understanding the global environment'.[18] The parties thus assume rather than explicitly state the key role of Antarctica in regulating different earth systems.

At the same time, from the beginning of the discussions, there was an acknowledgement that the protection of the continent could not be achieved in isolation from what happened beyond it and, vice versa, that caring for the world's environmental challenges could not be done without considering Antarctica. The Australia/France proposal made this point in a rather vague fashion:

> *The world community is recognizing that an integrated approach must be taken to environmental problems. Effects of human activity are no respecter of the artificial competences that may be set up between different institutions. A target obviously in need of protection may require attention to be given to a whole range of matters which may previously have been considered unrelated.*[19]

Peter Verbeek, head of the Dutch delegation to the first meeting in Viña del Mar, was much clearer about Antarctica's connections to the rest of the world and the implications of this for policymaking:

> *Damaging influences to the Antarctic environment also arise from outside Antarctica. The impact of human activities in Antarctica itself so far, fortunately, has been limited. The most important contribution to the objective of protecting the Antarctic environment, also therefore, should come from other*

[16] I understand 'systemic commons' as 'geographic areas of the earth that provide key water or ecosystem services, or help to regulate the climate system, so that their loss would greatly jeopardize the lives of human beings on earth': Alejandra Mancilla, 'Rethinking Land and Natural Resources, and Rights over Them', *Philosophy and Public Issues* 6, no. 2 (2016): 125–41, 138, https://images.fqpjournal.com/f/Archivio_rivista/FQ/FQP_2016_2. The concept of 'planetary commons' refers to 'globally shared geographic regions but also critical biophysical systems that regulate the resilience and state, and therefore livability, on Earth': Rockström et al., 'Planetary Commons'.

[17] Antarctic Treaty System, 'Protocol on Environmental Protection (1991)', Preamble.

[18] Ibid., Article III.

[19] Australia and France, 'Joint Australia/French Proposal', 4.

fora than this one. Our contribution here should be aimed as precisely and effectively as possible at minimizing adverse effects of the human presence in Antarctica itself.[20]

The fact that, to protect the Antarctic environment, decisions in other fora beyond the Antarctic Treaty were equally or even more relevant than the Treaty itself, was consistently ignored for the purposes of the Protocol. Instead, Verbeek's conclusion set the tone for what the Protocol eventually became: a legal tool for the regulation of human activities in the geographically demarcated area south of 60 degrees south. There was an irony in the fact that the Protocol's stated main goal was the 'comprehensive protection of the Antarctic environment and dependent and associated ecosystems'—a phrase that appeared fourteen times in the eighteen-page long document![21] Considering what was known already back then about global environmental connections, it is arguable that achieving this goal would have required the Environmental Protocol to be global in a double sense. First, insofar as its aim was the protection of the Antarctic environment, actions beyond Antarctica directly affecting Antarctica should have been regulated through it too. Second, because its aim was also the protection of dependent and associated ecosystems, the whole world—and not just Antarctica—should have become its object of concern. To give one concrete example, the future of the West Antarctic Ice Sheet is as dependent on global greenhouse gas emissions as the coastlines of Kiribati, Bangladesh, and the Netherlands are dependent on the future of the West Antarctic Ice Sheet. In practice, however, the interpretation of 'dependent and associated ecosystems' was much narrower, referring mostly to the sub-Antarctic area, and the measures listed to achieve comprehensive protection all referred to activities within the Antarctic Treaty area. The only hint at connecting the protection of Antarctica to activities happening beyond Antarctica came at the very end of the list, but with the aim of reporting rather than enforcing or regulating:

regular and effective monitoring shall take place to facilitate early detection of the possible unforeseen effects of activities carried on both within and outside the Antarctic Treaty area on the Antarctic environment and dependent and associated ecosystems.[22]

20 Republic of Chile, 'Interim Report', 54.
21 Antarctic Treaty System, 'Protocol on Environmental Protection (1991)'.
22 Ibid., Article III.

As Verbeek himself stated, this meant that the Protocol's goal was in effect to minimize disruptions to the environment by regulating human conduct in Antarctica. The most important goal—namely, to minimize disruptions to the Antarctic environment and to its dependent and associated ecosystems, no matter where these took place—was left for others to figure out. More than 30 years later, it remains unaddressed.

8.2 The problem of fit in Antarctica and how to address it

The dissonance between the intended goals of the Environmental Protocol and the measures proposed to achieve them gives rise to a 'problem of fit', defined as 'a matter of the match or congruence between biophysical systems and governance systems'.[23] In this case, there is an obvious *spatial misfit* between the area over which the Protocol rules and the area in which actions affecting Antarctica occur. The misfit is even larger if one considers the protection of Antarctica's dependent and associated ecosystems.[24] In other words, if the goal of the Protocol really was the protection of the Antarctic environment, as well as that of its dependent and associated ecosystems, then the geographical area over which the Protocol should have jurisdiction would have to be significantly extended. This is no news. As anthropologist Jessica O'Reilly reports after an encounter with a climate scientist frustrated by Antarctic environmental policy, 'he and his colleagues "don't care if the bases are tidy if the entire continent is starting to melt"'.[25] They regret that Antarctic environmental policy remains focused on practices to minimize human impact in Antarctica and is 'not far-sighted enough in terms of global change'.[26]

What is notorious in the Antarctic case is the utter disconnection between what the signatory states have committed themselves to doing (and refraining from doing) in Antarctica, and what they do (and refrain from doing) at home, despite the effects this might have for Antarctica. In 2020, 73 per cent of the CO_2 emissions and 64 per cent of the greenhouse gas emissions worldwide came from the twenty-nine consultative parties to the Antarctic Treaty, while seven out of the ten largest emitters were signatories to the Protocol.[27] This happens

[23] Young, 'Institutions and Environmental Change', 26.
[24] Ibid., 28. See also Arild Vatn and Paul Vedeld, 'Fit, Interplay, and Scale: A Diagnosis', *Ecology and Society* 17, no. 4 (2012): 12, http://dx.doi.org/10.5751/ES-05022-170412.
[25] O'Reilly, *Technocratic Antarctic*, 161.
[26] Ibid.
[27] The seven countries were China, the United States, India, Russia, Brazil, Japan, and Germany. The statistics are based on information provided in Hannah Ritchie, Pablo Rosado, and Max Roser,

at a time when it is no mystery that carbon emissions are the main drivers of global climate warming, which has momentous consequences for Antarctica. At the same time, these countries promise to accelerate the transition to the use of renewable energies in their Antarctic stations and even commit themselves to becoming carbon neutral.[28] For the sake of consistency, the least one could demand is that they also align their domestic environmental policies with the aims of the Protocol, so as not to flagrantly contradict these. In standard cases of spatial misfit, an institution might be unable to fulfil its goals because its jurisdiction is too small, or because it is 'unable to cope with actors or drivers external or internal and important for maintaining the ecosystem(s) or processes affected by the institution'.[29] In this case, instead, the members of the institution and the external actors are (almost) one and the same, but are acting in a disconnected way: chanting the importance of protecting the Antarctic environment during the Antarctic Treaty Meetings and other related fora, while being mostly oblivious to Antarctica when deciding on their environmental policies at home. If these dots were joined, the national environmental policies of the Treaty members would help rather than hinder the protection of the continent. This marks an important difference between the protection of Antarctica and the protection of other places obviously at risk from climate change. Small Island States and states in the Sahel region in Africa, for example, can impose strong mitigation measures in their territories and still be flooded by rising sea levels or doomed to suffer desertification. In their case, as opposed to Antarctica, it is not in their power to change the behaviour of others, even if it ends up affecting them negatively. Where in Antarctica the problem of fit can be addressed by requiring consistent action by member states, in this case those at risk can only invoke the goodwill of others. This should make the problem of fit in Antarctica, at least in this sense, easier to address. (Some might worry here that this would still be insufficient, given that more than three-quarters of world states are not bound by the Protocol. Thus, even if Antarctic states changed their conduct at home, this would not guarantee that third parties would not continue to behave in ways detrimental to the achievement of its

'Greenhouse Gas Emissions', Our World in Data, last updated January 2024, https://ourworldindata.org/greenhouse-gas-emissions.

[28] See Juan José Lucci, María Alegre, and Leandro Vigna, 'Renewables in Antarctica: An Assessment of Progress to Decarbonize the Energy Matrix of Research Facilities', *Antarctic Science* 34, no. 5 (2022): 374–88, https://doi.org/DOI: 10.1017/S095410202200030X; Kate Winter, 'Antarctica's First Zero Emission Research Station Shows That Sustainable Living Is Possible Anywhere', *The Conversation*, 19 November 2019, https://theconversation.com/antarcticas-first-zero-emission-research-station-shows-that-sustainable-living-is-possible-anywhere-113977.

[29] Vatn and Vedeld, 'Fit, Interplay, and Scale', Table 1.

goals. As I have mentioned, however, Antarctic Treaty members are together responsible for most global greenhouse gas emissions, and a change in their conduct could create pressure on others to change theirs too.)

That this disconnect in terms of acting persists until today is even more surprising considering that the causal interactions between Antarctica and the rest of the world have been abundantly documented and acknowledged. In 2021, 30 years after the signature of the Environmental Protocol, a group of Antarctic states finally recognized the profound effects that climate change was having on Antarctica, including 'the acceleration of ice flow and retreat in the Amundsen Sea Embayment of West Antarctica and in Wilkes Land, East Antarctica, which has the potential to lead to global mean sea-level rise of several metres within a few centuries'.[30] They also reported changes in ocean circulation, 'because of the significant amounts of human-produced heat and dissolved carbon in the Southern Ocean',[31] as well as 'changes in the distribution of krill, which will have consequential implications for the wider Antarctic ecosystem, including shifts in abundance and distribution of Antarctic mammals, birds and fish'.[32] In the preamble of a proposed draft resolution, moreover, the authors acknowledged the interconnection between a changing global climate and a changing Antarctica, declaring themselves '[*c*]*oncerned* at the implications to Antarctica from a warming global climate . . . including increasing melting of the West Antarctic Ice Sheet, and changes to Antarctic and Southern Ocean ecosystems' and '[*f*]*urther concerned* about the global implications of Antarctic change, notably sea-level rise and ocean circulation'.[33]

Despite this acknowledgement, however, they repeated—as if they were drawing from Verbeek's speech—that mitigating climate change was 'for another body', while the role of the Antarctic Treaty members was limited to 'researching and communicating Antarctic climate change science to the international audience'.[34] Concretely, they proposed that their governments' delegates to the Conference of the Parties of the United Nations Framework Convention on Climate Change (UNFCCC) 'are aware of the implications for Antarctica, unless action is taken to avoid dangerous climate change'; 'incorporate climate change considerations into the management of all human activities in Antarctica', and 'support their national Antarctic programmes and SCAR [the Scientific Committee on Antarctic Research] in their ongoing

[30] United Kingdom et al., 'Antarctica in a Changing Climate'.
[31] Ibid., 3.
[32] Ibid.
[33] Ibid., 4, their emphases.
[34] Ibid., 3.

efforts to undertake research about climate change and its impacts, and to communicate the implications for Antarctica, both within the Antarctic Treaty System and internationally'.[35] In practice, therefore, their proposal boiled down to bringing national delegates who are mindful of the effects of climate change on Antarctica to the key international meeting where this is discussed. Two years later, this draft evolved into the 'Helsinki Declaration on Climate Change and the Antarctic'.[36] Signed by the twenty-nine consultative parties of the Antarctic Treaty, it reiterated the previous diagnosis, '[r]eaffirming their firm commitment to combat the adverse impacts of climate change in Antarctica'.[37] In terms of actions, the declaration highlighted—among other things—the importance of continuing to do climate science in Antarctica and communicating its findings, giving support to the International Polar Year of 2032–33, celebrating the work done by the Committee for Environmental Protection, advising Antarctic operators on how to diminish their carbon footprint, and proposing to assess the risk posed by climate change to historic and heritage sites in Antarctica.[38] Among the eleven measures proposed, the Helsinki Declaration again failed to mention the most needed of them all: to wit, to align the domestic environmental policies of the signatories (especially those regarding greenhouse gas emissions) such that they promoted rather than detracted from the object and purpose of the Protocol. Considering that the members of the Antarctic Treaty represent around 63 per cent of the global population and include the G7 countries and the permanent members of the UN Security Council, making such a commitment would have had not just immediate effects but also immense value in terms of persuading others to join their quest. To invert the common saying, this was not like using a sledgehammer to crack a nut, but rather like using a nutcracker to demolish the Colosseum.

Some could point out that my assessment is too negative, and that the very fact that states have publicly recognized the challenges of a changing climate for Antarctica should be seen as progress. Moreover, some could invoke the notion of 'interplay' between and among global environmental governance institutions.[39] In line with the working paper mentioned above and with Verbeek's original statement, they could point out that it is fine to keep the Environmental Protocol as a legal instrument regulating human activity in

[35] Ibid., 4.
[36] Antarctic Treaty Meeting, 'Helsinki Declaration'.
[37] Ibid.
[38] Ibid.
[39] For the idea of interplay, see Young, 'Institutions and Environmental Change', 31–33.

Antarctica only, so long as the effects of global human activities to Antarctica are considered in other fora. This would be reflected when negotiating carbon emissions quota in the conferences of the parties of the UNFCCC, when setting biodiversity targets during the conferences of the parties of the Convention on Biodiversity, and so on. Ideally, one could imagine a scenario where national delegates managed to internalize or encapsulate Antarctica's interests to the point of making them their own during these negotiations.[40] As I discussed in previous chapters, however, it is hard to envisage an adequate representation of Antarctica by the states themselves. In cases of conflict, their short-term interests will weigh more heavily on national delegates than what is good for Antarctica, even if those short-term interests are detrimental to it. A more fitting solution, as mentioned in Chapters 5 and 6, would be to set up an independent body tasked with representing Antarctica directly in the relevant instances. But why would states allow or help to create such an institution if they find it so hard to represent Antarctica's interests in the first place? An analogy might help to answer this question. Humans worried about spending too much time on their smartphones install apps to limit and control their own use. Analogously here, states interested in the protection of Antarctica but aware of the danger of being oblivious to it in international negotiations could outsource this task to an independent entity. It is unlikely that this will be a top-down initiative taken by states, but it might happen if members of their own constituencies demand a change in this direction.

In sum, there are two moves that could diminish (if not fully resolve) the problem of fit at the heart of the Environmental Protocol. First, all member states should align their domestic environmental policies (to the extent that they are relevant for Antarctica) with the object and purpose of the Environmental Protocol. And second, they should ensure that Antarctica gets represented in the different fora of the Antarctic Treaty System, but also in other international meetings where decisions are made that have a momentous impact on the continent. While the former requires a more ambitious interpretation of the Environmental Protocol, the latter requires either a deep change in institutional culture (so that national delegates are ready to think of the interests of those beyond their constituencies) or an institutional reform, if one ultimately aims at the independent representation of Antarctica among other states.

[40] Recall the concept of internalization or encapsulation of interests in Goodin, 'Enfranchising the Earth.'

8.3 Does Antarctica have a right to be cold?

In *The Right to be Cold,* Inuit writer and activist Sheila Watt-Cloutier tells how a changing climate has impacted the life of her community and suggests that actions aiming at a stable climate, either through adaptation or mitigation, should be considered as required for the purpose of ensuring the protection of human rights.[41] In the case of the Inuit, the right to be cold is the right to continue pursuing their way of life without major disruptions. The failure of governments to attend to this right, by taking direct steps towards mitigation and adaptation, is in Watt-Cloutier's view a failure to protect the right of her people and many others whose lives will be shaken by the effects of climate change.

Does Antarctica have an analogous right to remain as it is without undergoing drastic change? Is that what would be required to protect it? In the previous section, I have made some assumptions in this direction. With the Environmental Protocol's object and purpose in mind, I have claimed that the comprehensive protection of Antarctica and dependent and associated ecosystems is desirable. Seeing how climate change is affecting Antarctica, I suggest that at least part of what it means to comprehensively protect the continent is to take measures to prevent it from melting. These measures would also contribute to the protection of Antarctica's associated and dependent ecosystems, which I take to be those low-lying coastal areas and islands around the world that will be most affected by the sea-level rise caused by the melting of Antarctic glaciers (especially the West Antarctic Ice Sheet).[42] Given that a straightforward measure to prevent Antarctica from melting is to drastically cut global greenhouse gas emissions, this is what the signatories of the Environmental Protocol ought to be doing.

Protecting Antarctica is desirable because it is a thriving community of life, a unique ecosystem, and a systemic commons or planetary commons with a key role in regulating the planet's temperature, air, and ocean currents. One could say here that all that matters is this second, instrumental function: so long as the continent continues to fulfil it, we should not worry too much about the extinction of particular species, or the disappearance of particular ecosystems. However, I have argued in previous chapters that we should move beyond

[41] Sheila Watt-Cloutier, *The Right to Be Cold: One Woman's Fight to Protect the Arctic and Save the Planet from Climate Change* (University of Minnesota Press, 2018), https://doi.org/10.5749/j.ctt2204r9f.

[42] Because the West Antarctic Ice Sheet floats in the sea, changes in temperature to the latter affect it directly. The case is different for the rest of the continent, which is land covered by ice and therefore not as affected by rising sea temperatures.

the assumptions of methodological anthropocentrism and start giving nature its due. Two authors have expressed this idea in illuminating ways. Edward L. McCord emphasizes the value of nature for us as valuers, and what is lost when other species are lost:

> *Human survival as a goal represents an exceedingly narrow ambition that is hardly commensurate with the full scope of human nature. Missed in the narrow calculus of human survival is the value of living things to curious and honest people for what they are in themselves. A wholesale destruction of other species in the family of life that has evolved over billions of years would represent an incalculable loss of the earth's heritage for mindful generations in the future.*[43]

In *Living on Earth*, Peter Godfrey-Smith worries about the erasing of life forms that have evolved in their own, unique ways, and that contribute to the richness of our planet:

> *When we preserve a diverse wild nature, we preserve, as far as we can, a great creative engine. This engine gave rise to us, along with other forms of life . . . We humans are a recent stem on the tree of animal evolution, one that has acquired the ability to look over the whole. There is a kind of ingratitude in allowing the extinctions, the narrowing, the reduction of animal life.*[44]

From an evolutionary perspective, Antarctica might not compare to other ecoregions in terms of diversity, but the adaptability of its inhabitants to such an extreme environment is itself worth preserving.

So, how should 'comprehensive protection' be understood? Like with other key terms in the document (e.g. intrinsic value), the term does not get defined in the Protocol but is left for parties to interpret. If the actual practices give us any hint, 'comprehensive protection' in Antarctica has been interpreted as the tight regulation of human activities happening in the area south of 60 degrees south, to the point where sometimes even rocks have to be put back in place (as in the case of the McMurdo Dry Valleys), and where any specimens taken have to be reported. One should not tread, but tiptoe, over Antarctica. 'Leave no trace' and 'Stay away from the natives' are the dicta that sum up this

[43] Edward L. McCord, *The Value of Species* (Yale University Press, 2012), 32–33.
[44] Peter Godfrey-Smith, *Living on Earth: Life, Consciousness and the Making of the Natural World* (William Collins, 2025), 247–48.

approach. But this should not make us forget one of the most important articles of the Protocol. At the time of its signature, the comprehensive protection of Antarctica probably alluded above all to Article 7, which banned mineral prospection and extraction. 'Comprehensive protection', in this context, was opposed to 'comprehensive exploitation'. Uniquely, Antarctica was turned into the only continent on earth where humans could not just extract resources at will. Protection happened *in situ*, and all the efforts of the member states were geared in this direction. In Chapter 7, I underscored this achievement and underlined the unique self-restraint displayed by states on this occasion, suggesting that applying that same spirit to rule over other places, like ecoregions, would be a welcome change of perspective.

This emphasis on the *in situ* dimension of protection neglected, however, the *ex situ* dimension, which is arguably equally important: to protect is 'to defend or guard from danger or injury; to support or assist against hostile or inimical action; to preserve from attack, persecution, harassment . . . to keep safe'.[45] To give an example, if we want to protect an organic farm from dangerous pollutants, it is not enough to ensure that the farm is kept clean; we also need to look around and see what the neighbours are doing (e.g. spraying Roundup upwind). One could claim, analogously, that just as an organic farm cannot prosper amid Roundupped fields, neither can Antarctica survive in a warming world. In the last two sections, when discussing what 'comprehensive protection' of Antarctica and its dependent and associated ecosystems required, I emphasized the importance of this *ex situ* dimension and underlined that the parties need to align their actions at home with the stated goals of the Protocol. I underlined moreover how in the Antarctic case one advantage is that the actors largely overlap: the same states that have pledged to protect Antarctica are the ones whose actions beyond Antarctica matter. To continue with the analogy above, this is a case where the owners of the organic farm are also the owners of most of the surrounding farms.

There is an ongoing discussion in ecology and environmental ethics regarding the normative value of 'stable' ecosystems, and the way in which we tend to normalize the environments that we know, turning them into reference points even though they have been shifting all along—what is known in the literature as the 'shifting baseline syndrome'.[46] In the case of Antarctica, one

[45] Oxford English Dictionary, 'protect (*v.*), sense 1.A', revised 2007, https://doi.org/10.1093/OED/4726075785.

[46] On the contested concept of ecosystem stability, see, for example, Jay Odenbaugh, 'Seeing the Forest and the Trees: Realism about Communities and Ecosystems', *Philosophy of Science* 74, no. 5 (2007): 628–41, https://doi.org/10.1086/525609; Mark Woods, *Rethinking Wilderness* (Broadview Press, 2017); and James Justus, *The Philosophy of Ecology: An Introduction* (Cambridge

could recall that, long before anthropogenic climate change became a concept, Antarctica had been changing since the dawn of time, and it had even been a rainforest 90 million years ago. Cycles of global warming and global cooling are not new to the continent. From a geological or deep-time perspective, it may sound both pointless and pompous to pretend that we cannot merely interfere with global climate, but that we can also steer it at will.

My response is that, even with the limited knowledge that we have of complex systems like the global climate, we can claim with relative certainty that greenhouse gas emissions contribute to global warming. This is expressed—among other ways—as ocean warming, which in turn contributes to the melting of Antarctic ice sheets (i.e. those parts of Antarctica that do not lie over land, but over water). This affects many Antarctic living communities and contributes to sea-level rise, which in turn deeply affects the inhabitants and the ecosystems of low-lying coastal areas and islands around the world. In other words, even if we might not be able to control the global climate at will, we do know what is causing it to become increasingly unstable and extreme, with areas like Greenland and Antarctica being the most obvious examples.

Consequently, 'to protect' should be understood in this context as doing what one can to keep a place safe from major disruptions that will not just be deleterious for some of those living in the place (the Antarctic community of living beings), but also for those threatened by changes in that place. States committed to protecting Antarctica have it in their power to cut their greenhouse gas emissions drastically, and this is what they should do. As mentioned earlier, some could object here that Antarctic states only represent a quarter of all states and that, even if they did act, this would not prevent non-members of the Treaty from continuing to burn fossil fuels. I have already shown, however, that Antarctic Treaty members produce the majority of CO_2 and greenhouse gas emissions, and they include all the major political, economic, and military powers. What they decide to do, therefore, would have a large impact on the global contribution to climate change and presumably on what others decide to do. Yet, some may worry that not all Antarctic states are equally prepared to undergo a fast transition to cleaner energies. Many of them are in fact developing countries that could end up harming their own citizens if they were to impose mitigation measures. To this I would reply that the principle

University Press, 2021). On the shifting baseline syndrome, see Masashi Soga and Kevin J. Gaston, 'Shifting Baseline Syndrome: Causes, Consequences, and Implications', *Frontiers in Ecology and the Environment* 16, no. 4 (2018): 222–30, https://doi.org/10.1002/fee.1794.

of 'Common but Differentiated Responsibilities' formalized by the UNFCCC could well be applied here too.[47] Even if not all of them should be expected to make the same effort, all should be expected to do something.

Some sceptics will point out at this stage that, even if all Antarctic states and all states cut their emissions drastically, this would still be no guarantee for the stability and continuity of the Antarctic ecosystem as we know it. Antarctica might still be doomed by external factors that are beyond anyone's control. But this argument deflects responsibility and turns the fact that we cannot control all variables into an excuse for not controlling even the one variable we know is making a difference. If states want to be trusted, they should align their actions with their speech. A future in which the West Antarctic Ice Sheet collapses in the face of states' inaction in tackling climate change is morally and politically gloomier than one where states have done what they could, with the same result ensuing.[48]

Summing up, one could argue that Antarctica has a right to be cold if by 'Antarctica' we understand the Antarctic community of life that will be severely affected by climate change. This right is also instrumental for Antarctic dependent and associated ecosystems, which will indirectly endure the effects of those changes. It might be true that Antarctic (and other) states can drastically cut their greenhouse gas emissions without this being a total guarantee that Antarctica will not undergo serious changes anyway. And yet this would fulfil what it means 'to protect' for humans and for human institutions: to do our best, based on the best available knowledge. Concentrating our efforts in protecting Antarctica in this way might not be sufficient to keep it as it is, but it is politically necessary. If not, it will be hard for states to avoid the charge that they are acting hypocritically and did not really mean what they signed up for.

[47] United Nations, 'United Nations Framework Convention on Climate Change', Preamble, 1992, https://unfccc.int/resource/docs/convkp/conveng.pdf.

[48] On this note, there have also been suggestions to slow down the melting of the glaciers by building 'sea curtains'. For a proposal to build an 80-kilometre-long curtain installed in 600-metre-deep waters in front of the Pine Island Glacier and Thwaites Glacier, see Bowie Keefer, Michael Wolovick, and John C. Moore, 'Feasibility of Ice Sheet Conservation Using Seabed Anchored Curtains', *PNAS Nexus* 2, no. 3 (2023): pgad053, https://doi.org/10.1093/pnasnexus/pgad053. See also John C. Moore, Rupert Gladstone, Thomas Zwinger, and Michael Wolovick, 'Geoengineer Polar Glaciers to Slow Sea-Level Rise', *Nature* 555, no. 7696 (2018): 303–5, https://doi.org/10.1038/d41586-018-03036-4. For a critical assessment from a governance perspective, see Patrick Flamm and Akiho Shibata, '"Ice Sheet Conservation" and International Discord: Governing (Potential) Glacial Geoengineering in Antarctica', *International Affairs* 101, no. 1 (2025): 309–20, https://doi.org/10.1093/ia/iiae281. Considering that, even if we reached net-zero emissions tomorrow, global warming will continue to happen for decades, these initiatives might be necessary too. The research is yet to be done, and one should adopt caution to avoid applying technofixes that could make things worse than what they already are.

8.4 Concluding remarks

In a provocative article, historian Peder Roberts adds to the existing critiques of science as the political currency in Antarctica.[49] In Roberts' view, anthropogenic climate change has turned Antarctica from being a closed laboratory for research to being a conduit through which the consequences of high CO_2 emissions are felt. At the same time, the inhabitants of Small Island States and other low-lying coastal areas—many of which are not members of the Antarctic Treaty—are some of the people in the world most affected by sea-level rise (which is significantly attributable to the melting of Antarctic ice). Furthermore, as mentioned already, some of the worst offenders in terms of global emissions are the consultative parties, whose failure to reduce their climate impact at home is turning Antarctica (especially the West Antarctic Ice Sheet) into a threat. In this scenario, would it not be fair to include those most affected by the melting of Antarctica at the decision-making table or even replace consultative parties with poor environmental records? Roberts is aware that adopting this *Most Affected Interests Principle* is more of a thought-provoking idea than a feasible proposal, but the point should be taken seriously.[50] As things stand, the very same parties that pledged to be the representatives of humankind for Antarctica are failing to live by their pledge: if they were genuinely interested in both the protection of Antarctica and the protection of those dependent on the protection of Antarctica, they would not keep contributing to its melting. Keeping science as the criterion for political participation in Antarctic affairs should thus require a minimal consistency from the governing countries, starting with them taking more aggressive measures to mitigate climate change at home as a condition for participation. More ambitiously (although I will not argue for that here), one could even suggest that the consultative parties should be especially responsible for helping with the relocation of the inhabitants of those countries that will be most affected by sea-level rise caused by the melting of Antarctic ice.[51] All this will sound overly naïve to many, but I would like to insist on the idea that we should uphold a normative vision of states as fiduciaries of humans and nature, rather than as self-interested agents pursuing the goals of a powerful few.

[49] Roberts, 'Science Criterion'.

[50] For a similar principle with a different scope of application, see Afsoun Afsahi, 'Towards a Principle of Most-Deeply Affected', *Philosophy & Social Criticism* 48, no. 1 (2022): 40–61, https://doi.org/10.1177/0191453720972740.

[51] I thank Andreas Føllesdal for suggesting this idea.

In this chapter, I have outlined the paradox of protection in Antarctica. Without doubt, the Environmental Protocol has been a force for good when it comes to the regulation of human activities and has enabled Antarctica to become the only continent where mineral extraction is banned. The Protocol, moreover, has allowed for the existence of Specially Managed Areas, Specially Protected Areas, and for specific rules geared at the protection of specific places. The idea that human interference should be kept to a minimum and that we should tread as little as possible has become widely accepted. At the same time, I have argued that a comprehensive protection of Antarctica and dependent and associated ecosystems requires not just action *in situ*, but also beyond Antarctic territory. The current misalignment between the actions of the member states in Antarctica and at home should be corrected, and the impact of climate change for Antarctica should be much more present not only in Antarctic fora, but also—and especially—in global climate negotiations.

While the slogan 'Think globally, act locally' entices people to consider the planet as a whole and to take local action accordingly, 'Think locally, act globally' would be a better dictum for the protection of Antarctica. By looking at what is done on the continent exclusively, Antarctic states have lost sight of the importance of global action for the preservation of Antarctica. Incorporating this consideration into their policies would not only help avert some catastrophic effects for Antarctic living communities and for communities whose future depends on Antarctica not melting, but it would also strengthen their position as genuine guardians of the place.

9
Governing Antarctica, Governing the World

> *Any discussion of the Antarctic problem, except perhaps in scientific circles, soon comes up hard on the question: ' ... but what's the use of it? What's the value of snow and ice so many miles away?' It is sometimes difficult to answer what earthly purpose the great white continent serves. It is not that the answer is necessarily lacking in logic or conviction; more often it is that the asking mind has not turned its thought in that direction; for the most unpractised student in polar history must soon sense its great significance. We lack, most of us, the universal, philosophical point of view. As is natural in our own crowded affairs, we see things narrowly, especially in a matter such as this, in trite personal and commercial terms of worth. Antarctica, 'a vast wonderland laid out on a giant scale, in which littleness has no place', cannot be judged, or appraised, according to limited values.*
>
> *Richard E. Byrd, Little America*[1]

We live in a world where the totality of the land surface (excluding Antarctica) is divided among states, which purport to have internal sovereignty over their 'own' affairs and external sovereignty understood as a right of noninterference by others. Fixing the political borders in pen, say its advocates, has diminished territorial conflicts and created a more peaceful world overall.[2] It has also allowed countries to develop economically, by exploiting their own natural resources, and it has given the different political constituencies the possibility to determine—or so it is hoped—their own future. At the same time, our prospects as humanity have never been as challenged, through a risky surpassing of most of our 'planetary boundaries' in the recent decades.[3] Anthropogenic

[1] Byrd, *Little America*, 38–39.

[2] See, for example, Wenar, *Blood Oil* and Steven Pinker, *The Better Angels of Our Nature: The Decline of Violence in History and Its Causes* (Penguin, 2011), 311–14.

[3] Katherine Richardson, Will Steffen, Wolfgang Lucht et al., 'Earth Beyond Six of Nine Planetary Boundaries', *Science Advances* 9, no. 37 (2023): eadh2458, https://doi.org/10.1126/sciadv.adh2458.

From Sovereignty to Guardianship. Alejandra Mancilla, Oxford University Press. © Alejandra Mancilla 2026.
DOI: 10.1093/9780197902769.003.0009

climate change, a steep increase in biodiversity loss, air and water pollution, the loss of healthy soils, and the introduction of new artificial substances are some of the boundaries that we have crossed that may lead us to unsafe, uncertain, and unpredictable developments. The emergence of our first global political and economic order has happened together with the emergence of environmental challenges that are to a large extent a consequence of that order and cannot be solved other than by political means. How are we to proceed?

In this concluding chapter, I suggest that we should look at Antarctica for guidance. In its 200 years as part of human history, Antarctica has been imagined as everything from the last treasure trove of resources for humanity to a green metropolis for sustainable growth to a sanatorium for the treatment of respiratory conditions to a nuclear waste dump. Instead, as explorer Richard E. Byrd proposes in the quote above, the real 'use' of Antarctica might be in forcing us to change our way of looking at things. To borrow the words of legal scholar Emilio Sahurie, especially when it comes to territorial governance, 'the political order of Antarctica may be its greatest legacy to mankind'.[4] As suggested in Chapter 7, some features of the Antarctic model (specifically its focus on guardianship of nature, through the joint cooperation of states) could be productively applied beyond Antarctica—for example, in places like ecoregions, the protection of which is key both to mitigate climate change and to avert further biodiversity loss. But, as Sahurie claims, the relevance of the Antarctic model might not be just in the model itself (which is, after all, a product of specific circumstances), but also in its value as a precedent that 'breaks with the fixity of legal concepts that obstruct both the intellectual process of clarification and the political process of implementation of common interests'.[5] In fact, this precedent of free thinking is probably the most valuable feature: a reminder that international law is an ongoing creation open to political imagination, and a reminder of the surprising results that can be obtained when states decide to work together rather than against each other.

I start by offering a brief overview of Antarctic futurology, a brimming field of speculation about the uses that Antarctica could eventually provide for humanity. I notice that a common feature of these imaginaries is that they all conceive of Antarctica as a resource for humans to benefit from. On the contrary, I suggest a different way to look at Antarctica, where its value is not as a material resource, but as a political source of inspiration. First, there are some aspects of the Antarctic model that may well be replicated (with adjustments) beyond

[4] Emilio J. Sahurie, *The International Law of Antarctica* (Martinus Nijhoff Publishers, 1992), 574.
[5] Ibid., 575.

the continent. Among these are the primacy of non-use, the protection of nature, the priority given to science, and the emphasis on cooperation between states. Second, the political will and legal imagination that made the Antarctic Treaty and its related instruments possible should give us hope that even more can be done; that the next step for the Antarctic Treaty System and territorial governance beyond Antarctica might be in recognizing that nature is not just the arena where human life unfolds, but something valuable for its own sake, such that it should be respected and taken into account in our political and legal frameworks. Given the tense geopolitics of our times and recent worrying developments at both the domestic and international levels, many speculate that Antarctica will sooner or later be subject to the same logic that we have applied everywhere else: territorial interests will be affirmed, minerals will be exploited, and the conflicts among world powers will be replicated on ice.[6] If this happens, has everything that I have said so far been in vain? I conclude by answering this question in the negative. For over 60 years, our nascent global order has managed to keep a whole continent apart, where states have used political imagination and shown self-restraint. Although sceptics might point out that this is explained by brute facts about Antarctica (too far, too cold, too dangerous, too expensive) rather than by good will, the result has nonetheless been remarkable and unique. Even if it has been *despite* themselves, states have treated Antarctica as a place to be protected and where precedence is given to priorities different from those that drive the rest of the world. The big hope of this book is that states end up protecting Antarctica and nature globally *because* it is the right thing to do.

9.1 The 'friendly' Antarctic

The Canadian-Icelandic explorer Vilhjalmur Stefansson became known in the early decades of the twentieth century for selling the idea of a 'friendly Arctic'.[7] Contrary to those who painted the Arctic regions as hostile to most humans, Stefansson proposed that they were actually inhabitable and productive year

[6] See, for example, Tiara Walters, "Krill, Baby, Krill'—What President-Elect Donald Trump Means for Imperilled Antarctica', *Daily Maverick*, 25 November 2024, https://www.dailymaverick.co.za/article/2024-11-25-krill-baby-krill-what-president-elect-donald-trump-means-for-imperilled-antarctica/ and Tiara Walters, 'Using Cape Town as a Launchpad, Russia Boasts of Supergiant Oil Fields in Antarctic Wilderness', *Daily Maverick*, 25 October 2021, https://www.dailymaverick.co.za/article/2021-10-25-using-cape-town-as-a-launchpad-russia-boasts-of-supergiant-oil-fields-in-antarctic-wilderness/.

[7] Vilhjalmur Stefansson, *The Friendly Arctic* (Macmillan Company, 1922).

round, if one had 'the right attitude'.[8] To put it in Lockean terms, the 'waste-land' could be turned into farmland so long as people were willing to mix their labour with the land. Stefansson's ideas were so influential that they helped to trigger an interest in the economic development of those regions and were the inspiration for some actual schemes in Canada and even Russia.

Nobody has echoed Stefansson's optimism about the Arctic when referring to Antarctica, but a string of authors at different times and places have wondered about its potential use for humans. In fact, it is as instructive to look at Antarctica's speculated futures as it is to look at its recorded past. What Antarctic futurologists share is a belief that Antarctica will be instrumental for humans, even if the time is not yet ripe to realize that value. Recall Douglas Mawson's perception of the frozen South: 'Who can say what a train of enterprise the future may bring?'[9]

Not surprisingly, the ideas of how to use Antarctica were influenced by what was going on in the rest of the world at the time. In the 1950s, Eric William Hunter Christie looked at the commercial potential of the continent, mainly through the whale and sea elephant industries. Especially the latter, he said, sounded promising, since 'it has been shown that sea-elephants, under proper control, may be farmed like cattle'.[10] He also proposed to import fur seals from the Falkland Islands and send Arctic foxes to the South Orkneys, and to bring Inuit as future colonizers, considering that the depletion of stocks and climate change in Greenland were already causing considerable suffering among those communities.[11] Frank Debenham, founder and director of the Scott Polar Research Institute in Cambridge, made a case for using Antarctica as the world's largest freezer: 'With the world population increasing at the present alarming rate there may well come a time when long term storage during periods of surplus productions will be necessary'.[12] At that point, Debenham remarked, Antarctica could become a long-term storage room with no running costs . . . except for transportation to and from the continent, which makes it 'hardly likely to be a commercial proposition for many years to come'.[13] Debenham also rightly guessed that more women would work on the continent. However, he did not envisage them in scientific stations, but rather in

[8] Howkins, *Polar Regions*, 101.

[9] Mawson, *Home of the Blizzard*, 6.

[10] Eric William Hunter Christie, *The Antarctic Problem* (George Allen & Unwin, 1951), 297.

[11] 'It is not perhaps too fantastic to consider the Eskimo as a future inhabitant of these regions, since, in parts of Greenland at any rate, climatic changes and the depletion of the stocks of native animals are causing a revolution in his habits, involving many of the smaller communities in considerable suffering': ibid.

[12] Frank Debenham, *Antarctica: The Story of a Continent* (Macmillan Company, 1961), 224–25.

[13] Ibid., 224.

large tourist resorts and sanatoria, where people with pulmonary diseases would travel to cure their ailments.[14]

The priorities changed during the 1970s, when—as shown in previous chapters—the exploitation of minerals became an active topic of discussion. The conjunction of the Middle Eastern water crisis together with the calving of a huge iceberg from the Larsen B ice shelf led more than one scientist to speculate about iceberg harvesting. In a report to the Australian Academy of Sciences, chemist Peter Schwerdtfeger concluded that:

> *In the light of potentially enormous rewards, both economic and humanitarian, a unique level of international cooperation to bring the concept of iceberg utilization to fruition in regions remote from Antarctica merits tangible encouragement and active participation by those nations which have relevant polar-operational and scientific and engineering expertise to contribute. It is at once significant and reassuring that icebergs constitute one of the very few, both Antarctic and global, naturally maintained, continuing resources.*[15]

In the 1980s, author Philip Quigg speculated about the possibility of turning Antarctica into a dump for radioactive waste:

> *if the costs were shared on a pro rata basis by all nations now seeking means of nuclear waste disposal, they might be manageable. Maintenance would be negligible, and the danger of sabotage would be minimal, although there would, of course, be hazards in shipping the material such long distances into a hostile environment.*[16]

While Quigg acknowledged that, at the time, the solution was merely 'academic', he thought it might eventually become 'the least unattractive alternative'.[17]

At the end of the 1980s, after nearly a decade of negotiations and when the consultative parties were putting the finishing touches on the Mineral Convention, legal scholar Oscar Pinochet de la Barra wrote that, in a 100 years' time, offshore oil in the hands of multinational consortia and atomic energy

[14] Ibid., 227.

[15] Peter Schwerdtfeger, 'On Icebergs and Their Uses', Report to the Australian Academy of Sciences, 1978, 46. For a historic overview of the idea of iceberg harvesting and recent attempts to resuscitate it, see Matthew H. Birkhold, *Chasing Icebergs: How Frozen Freshwater Can Save the Planet* (Pegasus Books, 2023).

[16] Philip W. Quigg, *A Pole Apart: The Emerging Issue of Antarctica* (New Press, 1983), 107.

[17] Ibid.

would be the powerhouses of Antarctica. While East Antarctica would remain the *far west* of the continent (forgive the pun), human settlements would take over the Peninsula and turn it into a lively metropole, home to over 10,000 inhabitants working in the krill-fishing and oil industries, tourism, and science.[18]

The most recent dream of an inhabited Antarctica has come from Doaa Abdel-Motaal. Writing after the 2015 refugee crisis in Europe, Abdel-Motaal has suggested that Antarctica could become a safe destination for climate refugees. Given that the climate in the Antarctic Peninsula is becoming like that of Greenland, one could think of a new Antarctic population settling there and eventually becoming their own, self-governing community. This would help reduce conflict among states currently dealing with ever-larger numbers of migrants seeking asylum.[19] As for the rest of the continent, the author imagines the claimants taking charge of their claimed slices and opening them for resource exploitation by private and public companies.[20]

This is far from an exhaustive review of Antarctic futurology, but it gives a flavour of the dominant view among authors and scholars imagining the future of the continent: a continent (mostly) without people, yet for people; a continent beyond states' territoriality, yet (mostly) run by states.[21] Methodological anthropocentrism and methodological nationalism thus prevail in these accounts. What I suggest next, instead, is that Antarctica's ultimate 'use' might not reside in its material value, but as a source of inspiration for global territorial governance; territorial governance that becomes mindful of both methodological anthropocentrism and nationalism and seeks to move beyond them.

9.2 Universalizing Antarctic values

Reflecting on the foundational values of the Antarctic Treaty System, Alan Hemmings mentions 'the preservation of peaceful purposes (for all intents and purposes meaning anything short of hostile military action); not rattling the bars of territorial claims; and the protection of scientific values'.[22] After the Environmental Protocol was signed, Hemmings adds, peace, science, and

[18] Oscar Pinochet de la Barra, *Antártica: Sueños del Ayer y del Mañana* (Hachette, 1988), 97–98. Ten pages on he concedes, however, that he dreams 'of the day when Antarctica will recover its freedom and will break the utilitarian chains that dominate the other continents': ibid., 118.

[19] Abdel-Motaal, *Antarctica*, 32.

[20] Ibid., 174.

[21] I say *mostly*, considering Pinochet's and Abdel-Motaal's proposals for what seem to be independent cosmopolitan communities. For a sci-fi version, see Kim Stanley Robinson, *Antarctica* (Harper Collins, 1997).

[22] Hemmings, 'Considerable Values', 143.

the protection of the environment became the 'new trinity of entrenched high values'.[23] Because these principles are increasingly threatened by the possibility 'to realize resource opportunities there, as everywhere else',[24] strengthening the commitment to them is as important as ever.

Hemmings is right that these are the cornerstones of Antarctic governance: the principles that get explicitly invoked in the Treaty and the Protocol, and the ones that states explicitly commit themselves to respecting when they become parties. They are also mentioned almost every time that Antarctica is described to a new audience as a one-of-a-kind territory, run by its own rules. The fact that these principles have been in place for over six decades has allowed for the development of a system of territorial governance with very special features. For the sake of keeping peace, the original claimant states, as well as newer ones with territorial aspirations, froze territorial sovereignty and embraced what I have called Joint Guardianship. If territorial rights elsewhere have been interpreted so far mostly as permission for states to exploit land and natural resources at their own discretion, in Antarctica parties are focused instead on their territorial duties: conducting science useful for understanding not just Antarctica, but the functioning of basic earth systems; protecting the place for present and future generations; turning preservation rather than exploitation into the default; and legally recognizing the value of nature independently of humans—something rare in international law. The suspension of the claims has thus allowed states to develop a cooperative and collaborative governance, and to change the focus from managing exploitation to managing preservation.

I concede that I am presenting the Antarctic Treaty System in an idealized way, and that—as Hemmings points out—these rules only hold for the land and not for the sea, where extraction has been the norm. Moreover, collaboration and cooperation have not been conflict-free, and consensus-based decision-making has made the system move very slowly (too slowly, some may say[25]). If states show an interest in Antarctica, sceptics will add, it is mere strategic thinking, gaining presence and getting ready for a time post-Treaty, when the pie will finally be divided.

Yet, I think one should resist the temptation of peeping into the 'real' intentions of states and look instead at what has been achieved even despite those intentions. For even if states agreed to this unique arrangement for purely

[23] Ibid., 145.
[24] Ibid., 153.
[25] See, for example, Nature Editorial, 'Reform the Antarctic Treaty', *Nature* 558, no. 14 (2018): 161, https://doi.org/10.1038/d41586-018-05368-7.

prudential reasons, we have still obtained for the most part a right result. Might this result be replicated beyond Antarctica, this time with a genuine motivation to play a fiduciary role? Here, I suggest taking a first step of political imagination and thinking of what our world would look like if the basic principles of international law were those of the Antarctic Treaty; if we turned peace, science, and the protection of the environment into the cornerstones of global territorial governance, while tempering the obsession of states with exclusive rights over 'their' land and natural resources.

As I showed in Chapters 2 and 3, the territorial claims in Antarctica were only comprehensible within the logic of late global colonialism, when states were dividing among themselves the last bits of *terra nullius*, even when it was full of people.[26] What in Antarctica look like evidently unwarranted claims to territory are thus not unique: newly independent countries around the same time were also engaging in the enlargement of their own territories, at the expense of Indigenous populations who had neither the institutional nor the military tools to fight back, or at the expense of nature, brought under someone's control when it could have just as well been left alone. Looking at the history of territorial appropriation of states should make us wary of grand claims of sovereignty over areas where other populations used to reside before, and over areas that were not humanly populated at all. This does not imply that states should revise their borders and shrink accordingly, but it does imply that the historicity and consequent fragility of their claims should be openly recognized. This would help, I think, in the orderly management of what I have called the systemic commons, areas that often spread over more than one state, but which need to be governed as units if we are to successfully preserve them.[27] Peaceful cooperation between states that share ecoregions would be the obvious place to start. If the relevant states agreed on limiting their right to exploit and focused instead on their duty to protect, this would go a long way to securing the future of the ecoregions. At the same time, because all states benefit from the existence of ecoregions, all states should cooperate in their preservation. This could be done by creating a fund to which all contribute

[26] See, for example, Angela Pratt, 'Treaties vs. Terra Nullius: Reconciliation, Treaty-Making and Indigenous Sovereignty in Australia and Canada', *Indigenous Law Journal* 3 (2004): 43–60, http://hdl.handle.net/1807/17116; Bruce Buchan and Mary Heath, 'Savagery and Civilization: From Terra Nullius to the "Tide of History"', *Ethnicities* 6, no. 1 (2006): 5–26, https://doi.org/10.1177/1468796806061077; and Andrew Fitzmaurice, 'The Genealogy of Terra Nullius', *Australian Historical Studies* 38, no. 129 (2007): 1–15, https://doi.org/10.1080/10314610708601228.

[27] For a proposal along the same lines, see Patrik Baard, 'Sovereignty, Ecology, and Regional Imperatives: Formulating Normative Foundations for Regional Ecological Justice', *Territory, Politics, Governance* (2024): 1–16, https://doi.org/10.1080/21622671.2023.2285302.

according to their means, by providing know-how and by conducting science in those ecoregions. Furthermore, in moving beyond methodological nationalism, the participation of local communities and nature representatives in the governance of these places should be taken seriously.

Beyond states' jurisdictions, a new governance model is also sorely needed. As Chris Armstrong has argued, this is obvious in the case of the oceans. As mentioned in Chapter 6, Armstrong proposes the creation of a World Ocean Authority to ensure that states focus on sustainable practices, leave a large percentage of the ocean protected, and incorporate coastal communities in the governance of their coastlines.[28] Moreover, Armstrong acknowledges the importance of giving certain rights to sentient marine animals, especially cetaceans.[29] To extend Armstrong's proposal, a World Land and Ocean Authority aimed at keeping peace, scientific cooperation, and the protection of the environment would signal a move into a new era of territorial governance. Not instead of, but in collaboration with states and other territorial actors, this would be the global body in charge of ensuring that our house (and not just its individual compartments) remains fit for habitation for all.

This may sound unrealizable today, but then so did the possibility of an Antarctic Treaty until just before its signature:

> *There is currently, a suggestion originating elsewhere but put forward by the President of the United States, that there should be a conference of the nations taking part in the I.G.Y. [International Geophysical Year] activities on the continent. The prime object of this conference would be to work out a formula under which the continent would be banned as a possible theatre of war. Such a pact would remove one of the reasons for distrust between the major nations who profess to have an interest in the Antarctic. It might come about if there were goodwill amongst all the nations concerned, but that is where we stumble on the essential difference between a family of children and a family of nations. The children have an all-powerful Papa who can enforce a decision but the family of nations has no Papa.*[30]

The same Frank Debenham who thought of Antarctica as the world's freezer was one of many who thought that states would never be able to contain their ambitions and cooperate. And yet it happened. Of course, that it happened in

[28] Armstrong, *Blue New Deal*, 207–22.
[29] Ibid., 137–57.
[30] Debenham, *Antarctica*, 239.

the past does not mean that it will happen again, but precedents give hope, especially when the other options seem untenable.

9.3 Universalizing Antarctic political imagination

I have claimed so far that Antarctic governance has been based on certain values and principles that may be fruitfully applied beyond Antarctica. In this section, I look at how critical and creative imagination in Antarctic governance should also serve as inspiration beyond it.

While 'rationality' and 'reason' figure prominently in political theory and political philosophy, much less attention has been paid to the imagination.[31] However, it is impossible to explain the development of international law (and of any law and social rule, for that matter) without appealing to this faculty. Avshalom M. Schwartz provides a definition of the political imagination and a tripartite division of it. A subset of the social imagination, political imagination has to do with the 'political institutionalization of social imaginaries' and with 'the concern of these imaginaries with power and power relations and with political practices such as legitimation and coercion'.[32] Schwartz proposes that there are three different types of imagination fulfilling different roles in the political realm. 'Constitutive' imagination is about keeping common codes and rules aiming at order and stability, to give the political order continuity over time. Any political order needs a shared imaginary, and this is what the constitutive imagination provides. 'Critical' imagination allows us to see that our social and political codes, norms, and institutions are constructed, historical, and open to problematization, reminding us of the possibility of deconstructing and reconstructing them. 'Creative' imagination, finally, allows for change and innovation, expands, challenges the status quo, and prevents political orders from becoming stagnant. Eventually, it may crystallize into constitutive imagination, to be questioned again.[33] Striking a balance between the three is the challenge and the ideal: stability and order are necessary for a political system to thrive, self-critique prevents it from becoming complacent and taking as given what is not, and creativity allows it to adapt and to change in response to the context.[34]

[31] Schwartz, 'Political Imagination'.
[32] Ibid., 3327.
[33] Ibid., 3328–29.
[34] Ibid., 3340–41.

One of the most remarkable features of Antarctic governance, in this sense, has been its use of critical and creative political imagination to constitute a one-of-a-kind arrangement. The very constitution of the Antarctic Treaty resulted from criticism against what was seen as an untenable status quo, premised upon the territorial claims of individual states. It established science for the first time as the main activity for a whole continent, it banned militarization and nuclearization, and it gave states—together—the responsibility to jointly administer a whole continent. I have mentioned these facts time and again through the book, but the point of repeating them here is to underline their existence as facts only because of human agreement.[35] The Antarctic Treaty System as it turned out to be was one of the many possible outcomes, a daring one that allowed for a considerable portion of the earth to be governed by a different logic from the rest of the world. After 30 years, when the idea of mineral exploitation began to be discussed, the workings of these two imaginations allowed the system to innovate again, while keeping its main features: by emphasizing the protection of the environment together with its other values, the commitment to keep Antarctica beyond the treadmill of sovereignty and exploitation was reinforced.

My claim here is that this peculiar Antarctic imagination is needed today to confront some of the political and environmental challenges that we are facing globally. There are three features that I wish to highlight. First, those creating the rules were able to recognize when a situation had no precedents and required new solutions. Suspending sovereignty was one such solution; drafting an Environmental Protocol was another. Second, Antarctic imagination has been a collective enterprise where states but also other actors have had a crucial role to play. One example is the influence that Greenpeace's idea for an Antarctic World Park had in modelling the objectives of the Environmental Protocol. Another are the contributions of the Scientific Committee on Antarctic Research (SCAR) to the discussions during the Antarctic Treaty Meetings. Third, the creation of Antarctic rules and institutions has not been excessively constrained (as elsewhere) by worries about what is politically feasible or not. This allowed, for example, for the imposition of a strict list of environmental measures that would have probably never been agreed upon individually by many of the signing states. Fourth, although still predominantly anthropocentric, Antarctic political imagination has like no other (excepting some non-Western approaches) incorporated nature as worthy of serious consideration.

[35] John Searle, *The Construction of Social Reality* (The Free Press, 1995).

To go back to the governance of ecoregions, applying these features of Antarctic imagination could be beneficial in the following ways. If states recognized that the protection of ecoregions required the collaboration not just of the states that have sovereignty over them but also of other states, then it would be possible to move to a new model of governance that more appropriately responds to what ecoregions require. Relatedly, this new model would come as a result of a collective deliberation not just by the community of states but also by other affected parties—most obviously among them, the human communities inhabiting these places. Moreover, those involved in the creation of a new governance system for ecoregions should ask themselves to what extent they are using arguments about political feasibility as conversation stoppers—with 'feasibility' masking lack of will to act rather than actual impossibility to act. Finally, nature should gain political and legal consideration. Bearing in mind that in ecoregions there already exists some level of collaboration and coordination between different territorial agents,[36] the openness to try Joint Guardianship that includes multiple actors—including nature representatives—may not be so far-fetched.

9.3.1 Isn't Antarctic governance itself in need of more Antarctic imagination?

Seen through the tripartite division of political imagination presented above, the Antarctic Treaty System today appears stagnant and not responsive enough to the internal and external challenges that it faces.[37] Among the internal challenges are changing global power balances and tensions (including a war) between member states, the difficulties of reaching consensus among twenty-nine consultative parties, and the inaction on topics that are discussed year after year without making any progress, like the regulation of tourism and bioprospecting. External challenges include an increase in illegal, unreported, and unregulated fishing, a latent interest in the possibilities of mineral extraction, and climate change. This is not an exhaustive list, but it gives a flavour

[36] See, for example, the Amazon Cooperation Treaty Organization (ACTO), formed by eight countries sharing the Amazon Forest: Amazon Cooperation Treaty Organization, 'About Us', https://otca.org/en/about-us/.

[37] See, for example, Alan D. Hemmings, 'Antarctic Politics in a Transforming Global Geopolitics', in *Handbook on the Politics of Antarctica*, ed. Klaus Dodds, Alan D. Hemmings, and Peder Roberts (Edward Elgar Publishing, 2017) and William Muntean, 'Considerations of Transparency in Antarctica', keynote speech at the seminar 'Strengthening the Transparency of the Antarctic Treaty System', Japan Society for the Promotion of Science, Antarctic Governance Research 2025–2030, 31 July 2025, https://youtu.be/apYfbhPJAzU.

of the issues that require action. So, isn't the Antarctic Treaty System itself in need of Antarctic imagination, particularly of the critical and creative kind? I think it is.

Critical imagination will be needed to revise the persistence of the sovereign claims that, while frozen, have not been given up and remain as relics of the colonial era; to rethink the rule of consensus, which might have been appropriate for twelve members, but has become increasingly paralyzing at twenty nine and counting; and to critically assess the activities of some national scientific programmes that resemble more an exercise in power play than an honest search for knowledge. Creative imagination will be required to devise new decision-making mechanisms if the rule of consensus is changed; to allow for the extension of the demos through a modified representation system; and to turn the Environmental Protocol into an instrument for the protection of the environment not just *in* Antarctica, but *for* Antarctica, by regulating the actions of its members not just in the continent but also beyond.

This might seem like a lot to ask. At the same time, the Antarctic Treaty—like no other international arrangement in the world—has been a counterexample to political realism in the international arena and to the claim that states only act to further their own selfish interests. While not denying that this has been part of the story, it has not been all of it. As Emilio Sahurie remarked in the early 1990s, just before the Environmental Protocol began to be discussed, appraising the Antarctic order must view its achievements 'in the perspective of the arms race, the disruption of global life support systems and environments, and the maintenance of a minimum public order'.[38] Member states today should revise these achievements and commit themselves to deepening them rather than giving them up.

9.4 What if?

In 2027, competition between the great powers over sourcing strategic minerals becomes intense due to their increased use in advanced weapons, communications equipment, and low carbon energy systems. China and Russia therefore encourage a southern African state, which is not party to the Antarctic Treaty or CCAMLR, to put forward plans for seabed exploration and mining of cobalt in the high seas areas below 60 degrees South . . . The US, Australia and many other 'likeminded' countries claim that, although this southern African

[38] Sahurie, *International Law*, 579.

> *country is not party to the Madrid Protocol, the mining prohibition is now customary in international law, so is binding on all states. The southern African state has refused to enter any arbitration or conciliation of the issue and is pressing ahead with a close logistics and military relationship with China and Russia, which plan to provide 'technology transfer' in the form of use of civilian ships to assist with transport of personnel and equipment for the seabed mining.*[39]

The scenario above is described as 'Cold War II' and is one of the five 'futures' speculated for Antarctica by political scientists Jeffrey McGee, David Edmiston, and Marcus Haward. More militarization, more spying on the neighbours, and an open race for valuable mineral resources needed for the 'green transition' set the tone here. For many in international relations and geopolitics, this is not just one *possible* future for Antarctica, but rather the *only* one conceivable. As I write these lines, one of the original signatories of the Antarctic Treaty, Russia, has invaded another country that is a consultative party to the Treaty, Ukraine. The rivalry between China and Australia in the Antarctic, and the fears of the West that China will become a polar power, make regular news.[40] There have been declarations by members of some scientific Antarctic national programmes quite openly admitting that they are interested in exploring the potential for future resource extraction.[41] Last, but not least, the erratic policy of the current US government, pulling out of international agreements and cutting science funding, are all worrying signs that the protective shield that has surrounded the continent until now is getting thinner and thinner.[42] So, what if it is? What if it is only a matter of time before Antarctica becomes yet another extractive site?

If this is the case, having written a whole book about possibilities for cooperation among humans based on an example that might cease to be an example could seem like a waste of time. In concluding, however, I would like to suggest that it is not. Hopefully, the example will endure, but even if it does not, the last

[39] McGee et al., *Future of Antarctica*, 159.

[40] See, for example, Elizabeth Buchanan, 'Great Power Competition Comes to Antarctica', *Foreign Affairs*, 18 March 2024, https://www.foreignaffairs.com/arctic-antarctic/great-power-competition-comes-antarctica.

[41] See, for example, Maya Maksimovich, 'Ukraine is Going to Compete for Its Leadership in Antarctica', *News of Ukraine*, 19 January 2019, https://replyua.net/129430-ukraina-sobiraetsya-sopernichat-za-pervenstvo-v-antarktide.html.

[42] While the United States has been very much at the centre of the discussions during the Antarctic Treaty Meetings, easily presenting twenty or more working papers and information papers in most years, in 2025 it presented only one: Antarctic Treaty Consultative Meeting, 'List of Documents', ATCM 47—CEP 2, Milan, 2025, https://documents.ats.aq/ATCM44/OP/ATCM44_OP009_e.pdf.

60 years of Antarctic governance would remain a beacon showing how politics can lead to remarkable results when the parties are willing to work together while keeping an open mind. Furthermore, Antarctic governance would remain a symbol of self-restraint, where political powers—even if grudgingly—put territorial aspirations to the side, focused on guardianship rather than use and exploitation of nature, and produced key scientific knowledge along the way. These two features (tight collaboration between political actors and self-restraint leading to a unique kind of territorial governance) might be the most valuable legacies of Antarctic governance to the world's political imaginary.

Interviewed during the Greenpeace campaign to turn Antarctica into a World Park at the end of the 1980s, activist Kelly Rigg pondered:

> *If we can't leave this place as it is—when there is no immediate need for its hidden mineral wealth, when there is sufficient time to find renewable alternatives, and when there is at least a degree of international cooperation in Antarctic science—how can we possibly hope to reverse the destructive tide in other areas of the world?*[43]

More than 30 years later, the need for minerals again seems more pressing (ironically this time for allowing the development of renewable sources of energy), and international cooperation remains limited. Yet, the Treaty endures. Turning around Rigg's rhetorical question, I ask: If we have managed to leave this place as it is—even when pressures to exploit rare minerals have grown, even when the global geopolitical scene is messy, and even when some states are still tempted to apply the standard territorial logic to Antarctica—shouldn't we hope to reverse the destructive tide in other areas of the world by bringing in the constructive Antarctic tide?

We have enough evidence that a system of territorial governance where the borders are historical rather than ecological is not serving us well. We have enough evidence that a global political order where states are the only actors leaves many of those relevantly affected out of the decision-making process. We have enough evidence that the default relationship between humans and their institutions on the one hand, and nature on the other, cannot keep being one of exploitation only. And we have evidence that fair and effective territorial governance can never be limited just to what is done in the territory itself; coordination between the actions of territorial actors is of the essence if the goal

[43] Quoted in John May, *The Greenpeace Book of Antarctica: A New View of the Seventh Continent* (Dorling Kindersley, 1988), 170.

is to keep our territories liveable, sustainable, and connected by relationships based on fairness. In this book, I have suggested that there is much that can be learnt from Antarctica in addressing these challenges. We should thus stop looking at Antarctica like an odd case study and regard it instead as a blueprint for a more adequate and effective form of territorial governance in this changing world.

Bibliography

Abdel-Motaal, Doaa. *Antarctica: The Battle for the Seventh Continent.* Praeger, 2016.

Abizadeh, Arash. 'Democratic Theory and Border Coercion: No Right to Unilaterally Control Your Own Borders'. *Political Theory* 36, no. 1 (2008): 37–65. http://journals.sagepub.com/doi/abs/10.1177/0090591707310090.

Adams, William M. 'Nature and the Colonial Mind'. In *Decolonizing Nature: Strategies for Conservation in a Post-Colonial Era*, edited by William M. Adams and Martin Mulligan. Earthscan, 2003. https://doi.org/10.4324/9781849770927.

Afsahi, Afsoun. 'Towards a Principle of Most-Deeply Affected'. *Philosophy & Social Criticism* 48, no. 1 (2022): 40–61. https://doi.org/10.1177/0191453720972740.

'Agreement on the Conservation of Albatrosses and Petrels (2001)'. Amended 7–11 May 2018. https://www.acap.aq/agreement-text/206-agreement-on-the-conservation-of-albatrosses-and-petrels/file.

Alliance of Small Island States (AOSIS). 'Alliance of Small Island States (AOSIS) Declaration on Climate Change 2009'. United Nations Sustainable Development, 21 September 2009. https://sustainabledevelopment.un.org/content/documents/1566AOSISSummitDeclarationSept21FINAL.pdf.

Amazon Cooperation Treaty Organization. 'About Us.' https://otca.org/en/about-us/.

Anderson, Atholl, Tipene O'Regan, Puamiria Parata-Goodall, Michael Stevens, and Te Maire Tau. 'On the Improbability of Pre-European Polynesian Voyages to Antarctica: A Response to Priscilla Wehi and Colleagues'. *Journal of the Royal Society of New Zealand* 52, no. 5 (2022): 599–605. https://doi.org/10.1080/03036758.2021.1973517.

'Annual Precipitation'. Our World in Data, 2024. https://ourworldindata.org/grapher/average-precipitation-per-year.

Antarctic Rights. 'Working Group'. 2025. https://antarcticrights.org/about/members/.

Antarctic Treaty Consultative Meeting. 'List of Documents'. ATCM 47—CEP 2, Milan, 2025. https://documents.ats.aq/ATCM44/OP/ATCM44_OP009_e.pdf.

Antarctic Treaty Meeting. 'Helsinki Declaration on Climate Change and the Antarctic, Resolution 2'. ATCM XLV—CEP XXV, Helsinki, 2023. https://documents.ats.aq/ATCM45/ad/atcm45_ad006_e.pdf.

Antarctic Treaty System. 'Convention for the Conservation of Antarctic Seals (1972)'. https://documents.ats.aq/keydocs/vol_1/vol1_13_CCAS_CCAS_e.pdf.

Antarctic Treaty System. 'Annex II to the Protocol on Environmental Protection to the Antarctic Treaty: Conservation of Antarctic Fauna and Flora (1991)'. https://documents.ats.aq/recatt/Att432_e.pdf.

Antarctic Treaty System. 'Protocol on Environmental Protection to the Antarctic Treaty (1991)'. https://documents.ats.aq/recatt/Att006_e.pdf.

Aristotle. 'Metaphysics'. In *The Complete Works of Aristotle, Bollingen Series LXXI-2*, edited by Jonathan Barnes. Princeton University Press, 1984.

Aristotle. 'Meteorology'. In *The Complete Works of Aristotle, Bollingen Series LXXI-2*, edited by Jonathan Barnes. Princeton University Press, 1984.

Armstrong, Chris. 'Against "Permanent Sovereignty" over Natural Resources'. *Politics, Philosophy & Economics* 14, no. 2 (2015): 129–51. http://journals.sagepub.com/doi/abs/10.1177/1470594X14523080.

Armstrong, Chris. 'Sharing Conservation Burdens Fairly'. *Conservation Biology* 33, no. 3 (2019): 554–60. https://doi.org/10.1111/cobi.13260.

Armstrong, Chris. *A Blue New Deal: Why We Need a New Politics for the Ocean.* Yale University Press, 2022.

Arneil, Barbara. *John Locke and America: The Defence of English Colonialism*. Clarendon Press, 1996.

Arneil, Barbara. 'Colonialism Versus Imperialism'. *Political Theory* 52, no. 1 (2024): 146–76. https://doi.org/10.1177/00905917231193107.

Associated Press. 'Coldest Temperature Ever Recorded on Earth in Antarctica: −94.7C (−135.8F)'. *The Guardian*, 10 December 2013. https://www.theguardian.com/world/2013/dec/10/coldest-temperature-recorded-earth-antarctica-guinness-book.

Auburn, F. M. 'Antarctic Minerals and the Third World'. *FRAM: The Journal of Polar Studies* 1 (1984): 200–23.

Australia and France. 'A Joint Australia/French Proposal in the Form of a Paper Including a Draft Recommendation for ATCM XV: Comprehensive Measures for the Protection of the Antarctic Environment and Its Dependent and Associated Ecosystems'. XV ATCM/WP2, Paris, 9–19 October 1989. https://documents.ats.aq/ATCM15/wp/ATCM15_wp002_e.pdf.

Australia and USA. 'Conservation at CCAMLR: Understanding Article II of the Convention on the Conservation of Antarctic Marine Living Resources'. CCAMLR-XXXV/BG/28, 2016. https://www.antarctica.gov.au/site/assets/files/47746/camlr-art-ii-interpretation-final.pdf.

Australian Antarctic Program. 'Flying Birds'. Last updated 12 August 2010. https://www.antarctica.gov.au/about-antarctica/animals/flying-birds/.

Australian Antarctic Program. 'Australian Antarctic Territory'. Last updated 15 April 2016. https://www.antarctica.gov.au/about-antarctica/australia-in-antarctica/australian-antarctic-territory/.

Australian Antarctic Program. 'Antarctic Krill'. Last updated 28 February 2025. https://www.antarctica.gov.au/about-antarctica/animals/krill/.

Australian Government, Federal Register of Legislation. *Australian Antarctic Territory Acceptance Act (1933)*. Latest version 31 December 1973. https://www.legislation.gov.au/C1933A00008/latest/text.

Australian Government (Geoscience Australia). 'Australian Antarctic Territory'. Last updated 30 November 2022. https://www.ga.gov.au/scientific-topics/national-location-information/dimensions/remote-offshore-territories/australian-antarctic-territory.

Baard, Patrik. 'Sovereignty, Ecology, and Regional Imperatives: Formulating Normative Foundations for Regional Ecological Justice'. *Territory, Politics, Governance* (2024): 1–16. https://doi.org/10.1080/21622671.2023.2285302.

Barros Recabarren, Manuel. 'Presentación'. In *¡Base O'Higgins Sin Novedad!*. Second edition, by Hugo Schmidt Prado. Editorial La Noria, 1992.

Bastmeijer, Kees. 'The Antarctic Treaty Turning Sixty: Does Aging Affect Decision-Making? The Case of Antarctic Tourism'. Presentation at the workshop 'On Its 60th Birthday, Is the Antarctic Treaty in Good Health?', organized by the project 'Political Philosophy Looks to Antarctica'. University of Oslo, September 2019.

Bastmeijer, Kees, Akiho Shibata, Imme Steinhage, Luis Valentín Ferrada, and Evan T. Bloom. 'Regulating Antarctic Tourism: The Challenge of Consensus-Based Decision Making'. *American Journal of International Law* 117, no. 4 (2023): 651–76.

Beck, Peter J. *The International Politics of Antarctica*. Croom Helm, 1986.

Beck, Peter J. 'The United Nations and Antarctica, 2002: "Constructive Engagement" Between the Antarctic Treaty System and the United Nations'. *Polar Record* 39, no. 3 (2003): 231–38. https://doi.org/DOI: 10.1017/S0032247403003103.

Beck, Peter J. 'Twenty Years On: The UN and the "Question of Antarctica," 1983–2003'. *Polar Record* 40, no. 3 (2004): 205–12. https://doi.org/DOI: 10.1017/S0032247403003371.

Beck, Peter J. 'The United Nations and Antarctica, 2005: The End of the "Question of Antarctica"'? *Polar Record* 42, no. 3 (2006): 217–27. https://doi.org/10.1017/S003224740600533X.

Benvenisti, Eyal. 'Sovereigns as Trustees of Humanity: On the Accountability of States to Foreign Stakeholders'. *American Journal of International Law* 107, no. 2 (2013): 295–333. https://doi.org/10.5305/amerjintelaw.107.2.0295.

Bergami, E., C. Manno, S. Cappello, M. L. Vannuccini, and I. Corsi. 'Nanoplastics Affect Moulting and Faecal Pellet Sinking in Antarctic Krill (*Euphausia Superba*) Juveniles'. *Environment International* 143 (2020): 105999. https://doi.org/10.1016/j.envint.2020.105999.

Berguño Barnes, Jorge. 'Historia Intelectual del Tratado Antártico'. *Boletín Antártico Chileno* 19, no. 1 (2000): 2–12.

Birkhold, Matthew H. *Chasing Icebergs: How Frozen Freshwater Can Save the Planet.* Pegasus Books, 2023.
Boatner, Kay. 'Tardigrade'. *National Geographic Kids.* https://kids.nationalgeographic.com/animals/invertebrates/facts/tardigrade.
Borrows, John. *Law's Indigenous Ethics.* University of Toronto Press, 2019.
Boyd, David R. *The Rights of Nature: A Legal Revolution That Could Save the World.* ECW Press, 2017.
Brady, Anne-Marie. *China as a Great Polar Power.* Cambridge University Press, 2017.
Braun, Christina, Fritz Hertel, and Hans-Ulrich Peter. 'Environmental Management: The Fildes Peninsula Paradigm'. In *Handbook on the Politics of Antarctica*, edited by Klaus Dodds, Alan D. Hemmings, and Peder Roberts. Edward Elgar Publishing, 2017.
Brazilian Coalition on Climate, Forests and Agriculture. 'About Us'. https://coalizaobr.com.br/en/about-us/.
British Antarctic Survey. 'Highly Pathogenic Avian Influenza (HPAI) Has Been Confirmed in Brown Skua Populations on Bird Island, South Georgia—The First Known Cases in the Antarctic Region'. 23 October 2023. https://www.bas.ac.uk/media-post/first-confirmed-cases-of-avian-influenza-in-the-antarctic-region/.
Brooks, Shaun, and Julia Jabour. 'Australia Wants to Build a Huge Concrete Runway in Antarctica. Here's Why That's a Bad Idea'. *The Conversation*, 17 July 2020. https://theconversation.com/australia-wants-to-build-a-huge-concrete-runway-in-antarctica-heres-why-thats-a-bad-idea-139596.
Brownlie, Ian. *Principles of International Law.* Seventh edition. Oxford University Press, 2008.
Buchan, Bruce, and Mary Heath. 'Savagery and Civilization: From Terra Nullius to the "Tide of History"'. *Ethnicities* 6, no. 1 (2006): 5–26. https://doi.org/10.1177/1468796806061077.
Buchanan, Allen. *Justice, Legitimacy and Self-Determination: Moral Foundations for International Law.* Oxford University Press, 2004.
Buchanan, Elizabeth. 'Great Power Competition Comes to Antarctica'. *Foreign Affairs*, 18 March 2024. https://www.foreignaffairs.com/arctic-antarctic/great-power-competition-comes-antarctica.
Bufacchi, Vittorio. 'Colonialism, Injustice, and Arbitrariness'. *Journal of Social Philosophy* 48, no. 2 (2017): 197–211. https://doi.org/10.1111/josp.12182.
Bush, Barbara. *Imperialism and Postcolonialism.* Routledge, 2014.
Butt, Daniel. 'Colonialism and Postcolonialism'. In *The International Encyclopaedia of Ethics*, edited by Hugh LaFollette. Wiley-Blackwell, 2013.
Byrd, Richard Evelyn. *Little America: Aerial Exploration in the Antarctic and the Flight to the South Pole.* G. P. Putnam's Sons, 1930.
Callicott, J. Baird. 'Animal Liberation: A Triangular Affair'. *Environmental Ethics*, no. 4 (1980): 311–38.
Callicott, J. Baird. *Thinking Like a Planet: The Land Ethic and the Earth Ethic.* Oxford University Press, 2013.
Campaign for Nature. 'G7 Leaders Agree to Historic "Nature Compact"—Set Comprehensive Biodiversity Targets, Commit to Protecting At Least 30% of Lands and Seas'. Media statement. 13 June 2021. https://www.campaignfornature.org/g7-nations-agree-to-historic-nature-compact.
Caney, Simon. 'Democratic Reform, Intergenerational Justice and the Challenges of the Long-Term'. Centre for the Understanding of Sustainable Prosperity, 2019. https://www.cusp.ac.uk/wp-content/uploads/11-Simon-Caney-online.pdf.
Cardone, Ignacio Javier. 'Argentina and Chile's Antarctic Colonialism? A Postcolonial Critique to Eurocentric Analysis'. In *Colonialism and Antarctica: Attitudes, Logics, and Practices*, edited by Peder Roberts and Alejandra Mancilla. Manchester University Press, 2024. https://www.manchesterhive.com/display/9781526170644/9781526170644.00011.xml.
Celermajer, Danielle, David Schlosberg, Lauren Rickards, et al. 'Multispecies Justice: Theories, Challenges, and a Research Agenda for Environmental Politics'. *Environmental Politics* 30 (2021): 119–40. https://doi.org/10.1080/09644016.2020.1827608.
Césaire, Aimé. *Discourse on Colonialism.* Translated by Joan Pinkham. Monthly Review Press, 2000.
Chapple, Geoff. 'Harvest of Souls'. *New Zealand Geographic*, July–August 2005. https://www.nzgeo.com/stories/harvest-of-souls/.
Chaturvedi, Sanjay. *The Polar Regions: A Political Geography.* Wiley, 1996.

Chaturvedi, Sanjay. 'Rise and Decline of Antarctica in Nehru's Geopolitical Vision: Challenges and Opportunities of the 1950s'. *The Polar Journal* 3, no. 2 (2013): 301–15. https://doi.org/10.1080/2154896X.2013.868087.

Cherry-Garrard, Apsley. *The Worst Journey in the World*. Basic Books, 2008.

Christie, Eric William Hunter. *The Antarctic Problem*. George Allen & Unwin, 1951.

Cochrane, Alasdair. 'Cosmozoopolis: The Case Against Group-Differentiated Animal Rights'. *Law, Ethics and Philosophy* no. 1 (2013): 127–41. http://www.raco.cat/index.php/LEAP/article/view/294785.

Cochrane, Alasdair. *Sentientist Politics: A Theory of Global Inter-Species Justice*. Oxford University Press, 2018.

Cochrane, Alasdair, Robert Garner, and Siobhan O'Sullivan. 'Animal Ethics and the Political'. *Critical Review of International Social and Political Philosophy* 21, no. 2 (2018): 261–77. https://www.tandfonline.com/action/showCitFormats?doi=10.1080/13698230.2016.1194583.

Coleridge, Samuel Taylor. *Selected Poetry and Prose*. Holt, Rinehart and Winston, 1951.

Comisión Nacional del Antártico. *Soberanía Argentina en la Antártida*. Second edition. Ministerio de Relaciones Exteriores y Culto, 1948.

Commission for the Conservation of Antarctic Marine Living Resources. 'Convention for the Conservation of Antarctic Marine Living Resources'. 1982. Page last modified 25 November 2013. https://www.ccamlr.org/en/organisation/camlr-convention-text.

Conference on Antarctica. 'The Antarctic Treaty'. 1 December 1959. http://www.ats.aq/documents/ats/treaty_original.pdf.

Constitución de la República del Ecuador 2008. Last modified 13 July 2011. https://www.oas.org/juridico/pdfs/mesicic4_ecu_const.pdf.

Convey, Peter, and Sandra J. McInnes. 'Exceptional Tardigrade-Dominated Ecosystems in Ellsworth Land, Antarctica'. *Ecology* 86, no. 2 (2005): 519–27. https://doi.org/https://doi.org/10.1890/04-0684.

Convey, Peter, and Lloyd S. Peck. 'Antarctic Environmental Change and Biological Responses'. *Science Advances* 5, no. 11 (2019): eaaz0888. https://doi.org/10.1126/sciadv.aaz0888.

Corrigan, Daniel P., and Markku Oksanen, eds. *Rights of Nature: A Re-Examination*. Routledge, 2021. https://doi.org/10.4324/9780367479589.

Costanza, Robert, Ralph D'Arge, Rudolf De Groot, et al. 'The Value of the World's Ecosystem Services and Natural Capital'. *Nature* 387, no. 6630 (1997): 253–60. https://www.nature.com/articles/387253a0.pdf.

Council of Europe. 'Convention on the Conservation of European Wildlife and Natural Habitats (ETS no. 104)'. 19 September 1979. https://rm.coe.int/1680078aff.

Criddle, Evan J., and Evan Fox-Decent. *Fiduciaries of Humanity: How International Law Constitutes Authority*. Oxford University Press, 2016.

Cripps, Elizabeth. *Climate Change and the Moral Agent: Individual Duties in an Interdependent World*. Oxford University Press, 2013. https://doi.org/10.1093/acprof:oso/9780199665655.001.0001.

Cruz, María Jimena, Melisa A. Salerno, and Andrés Zarankin. 'Nineteenth-Century Connections Between Capitalism and Colonialism in Antarctica: The Case of Sealing in the South Shetlands'. In *Colonialism and Antarctica: Attitudes, Logics, and Practices*, edited by Peder Roberts and Alejandra Mancilla. Manchester University Press, 2024. https://doi.org/10.7765/9781526170644.00015.

Cullinan, Cormac. *Wild Law: A Manifesto for Earth Justice*. Second edition. Green Books, 2011.

Cullinan, Cormac. 'The Antarctic Realm: Exploring the Potential of Antarctica as a Sovereign Entity'. Presentation at the workshop 'Antarctica and Rights of Nature: Historical, Moral and Legal Perspectives', organized by the projects 'Greening the Poles' and 'Dynamic Territory'. Lysebu, August 2022.

Curley, Andrew. 'Resources Is Just Another Word for Colonialism'. In *The Routledge Handbook of Critical Resource Geography*, edited by Matthew Himley, Elizabeth Havice, and Gabriela Valdivia. Taylor and Francis, 2021.

Dahl, Robert A. *After the Revolution?: Authority in a Good Society*. Yale University Press, 1990.

Dalrymple, William. *The Anarchy: The Relentless Rise of the East India Company*. Bloomsbury Publishing, 2019.

Dasgupta, Partha. 'The Economics of Biodiversity: The Dasgupta Review—Headline Messages'. UK Government Publishing Service, February 2021. https://assets.publishing.service.gov.uk/media/60182857d3bf7f70c2afe5bb/Dasgupta_Review_-_Headline_Messages.pdf.
Dasgupta, Partha. *The Economics of Biodiversity*. Cambridge University Press, 2024. https://doi-org.ezproxy.uio.no/10.1017/9781009494359.
Daverede, Alberto L. 'Política y Actividades Antárticas de La República Argentina'. In *Aula de Estudios Antárticos*. Madrid, 1987.
Debenham, Frank. *Antarctica: The Story of a Continent*. Macmillan Company, 1961.
DeMarle, David J. 'Design Parameters for a South African Iceberg Power and Water Project'. In *Proceedings of the Conference on Use of Icebergs: Scientific and Practical Feasibility, Cambridge, UK, 1–3 April 1980*. International Glaciological Society, 1980, 129–133.
Dinerstein, E., A. R. Joshi, C. Vynne, et al. 'A "Global Safety Net" to Reverse Biodiversity Loss and Stabilize Earth's Climate'. *Science Advances* 6, no. 36 (2020): eabb2824. https://doi.org/10.1126/sciadv.abb2824.
Dodds, Klaus. *Pink Ice: Britain and the South Atlantic Empire*. I. B. Tauris & Co., 2002.
Dodds, Klaus. 'Post-Colonial Antarctica: An Emerging Engagement'. *Polar Record* 42, no. 1 (2006): 59–70.
Dodds, Klaus. *The Antarctic: A Very Short Introduction*. Oxford University Press, 2012.
Dodds, Klaus. 'Assault on the Unknown: Geopolitics, Antarctic Science and the International Geophysical Year (1957–8)'. In *New Spaces of Exploration: Geographies of Discovery in the Twentieth Century*, edited by Simon Naylor and James R. Ryan. I. B. Tauris, 2015.
Donaldson, Sue, and Will Kymlicka. *Zoopolis: A Political Theory of Animal Rights*. Oxford University Press, 2011.
Donaldson, Sue, and Will Kymlicka. 'A Defense of Animal Citizens and Sovereigns'. *Law, Ethics and Philosophy* no. 1 (2013): 143–60. https://raco.cat/index.php/LEAP/article/view/294786.
Donoso, Alfonso. 'Valor intrínseco en la Antártica'. *Revista de Estudios Hemisféricos y Polares* 12, no. 1 (2021): 44–57.
Donoso, Alfonso. 'Intrinsic Value, Sovereignty Rights, and Sentient Animals in Antarctica'. *The Polar Journal* 14, no. 2 (2024): 373–88. https://doi.org/10.1080/2154896X.2024.2415809.
Dryzek, John S. *The Politics of the Earth: Environmental Discourses*. Fourth edition. Oxford University Press, 2022.
Dryzek, John S., and Simon Niemayer. 'Discursive Representation'. *American Political Science Review* 102, no. 4 (2008): 481–93. https://doi.org/DOI: 10.1017/S0003055408080325.
Dudeney, John R., and David W. H. Walton. 'Leadership in Politics and Science Within the Antarctic Treaty'. *Polar Research* 31 (2012): 1–9. https://doi.org/10.3402/polar.v31i0.11075.
Duffy, Rosaleen. *Nature Crime: How We Are Getting Conservation Wrong*. Yale University Press, 2010.
Dumont d'Urville, Jules Sébastien César. *An Account in Two Volumes of Two Voyages to the South Seas by Captain Jules S-C. Dumont d'Urville of the French Navy to Australia, New Zealand, Oceania 1826–1829 in the Corvette Astrolabe, and to the Straits of Magellan, Chile, Oceania, South East Asia, A*. Vol. 2. Translated and edited by Helen Rosenman. Melbourne University Press, 1987.
Eckersley, Robyn. *The Green State: Rethinking Democracy and Sovereignty*. MIT Press, 2004.
Eckersley, Robyn. 'Representing Nature'. In *The Future of Representative Democracy*, edited by John Keane, Sonia Alonso, and Wolfgang Merkel. Cambridge University Press, 2011. https://doi.org/DOI: 10.1017/CBO9780511770883.011.
Eijgelaar, Eke, Carla Thaper, and Paul Peeters. 'Antarctic Cruise Tourism: The Paradoxes of Ambassadorship, "Last Chance Tourism" and Greenhouse Gas Emissions'. *Journal of Sustainable Tourism* 18, no. 3 (2010): 337–54. https://doi.org/10.1080/09669581003653534.
'Elephants' Graveyard No More'. *The Economist*, 24 October 2020, 31–32.
Elzinga, Aant. 'Antarctica: The Construction of a Continent by and for Science'. In *Denationalizing Science*, edited by Elisabeth Crawford, Terry Shinn, and R. D. Whitley. Springer, 1993. https://doi.org/10.1007/978-94-017-1221-7_3.
Elzinga, Aant. 'Geopolitics, Science and Internationalism During and After IGY'. In *Boletín Antártico Chileno: Second SCAR Workshop on the History of Antarctic Research (Santiago, Chile, 21 to 22 September* 2006*)*, edited by Reiner Canales Cabezas. La Prensa Austral, 2006. https://www.inach.cl/wp-content/uploads/2009/10/boletin-historico-agosto-2009.pdf.

Elzinga, Aant. 'The Continent for Science'. In *Handbook on the Politics of Antarctica*, edited by Klaus Dodds, Alan Hemmings, and Peder Roberts. Edward Elgar Publishing, 2017.

Escobar, Arturo. 'Degrowth, Postdevelopment, and Transitions: A Preliminary Conversation'. *Sustainability Science* 10, no. 3 (2015): 451–62. https://doi.org/10.1007/s11625-015-0297-5.

Fedorov, Yevgeny K. 'Antarctica: Experimental Proving Ground for Peaceful Coexistence and International Collaboration'. In *Frozen Future*, edited by Richard S. Lewis and Philip M. Smith. Quadrangle Books, 1973.

Fitzmaurice, Andrew. 'The Genealogy of Terra Nullius'. *Australian Historical Studies* 38, no. 129 (2007): 1–15. https://doi.org/10.1080/10314610708601228.

Flamm, Patrick. 'Legitimating the Antarctic Treaty System: From Rich Nations Club to Planetary Ecological Democracy?' *Australian Journal of International Affairs* 76, no. 3 (2022): 266–85. https://doi.org/10.1080/10357718.2022.2056876.

Flamm, Patrick, and Akiho Shibata. '"Ice Sheet Conservation" and International Discord: Governing (Potential) Glacial Geoengineering in Antarctica'. *International Affairs* 101, no. 1 (2025): 309–20. https://doi.org/10.1093/ia/iiae281.

Fontana, Pablo. *La Pugna Antártica: El Conflicto por el Sexto Continente 1939–1959*. Guazuvirá Ediciones, 2014.

Franks, Becca, Christopher Ewell, and Jennifer Jacquet. 'Animal Welfare Risks of Global Aquaculture'. *Science Advances* 7, no. 14 (2023): eabg0677. https://doi.org/10.1126/sciadv.abg0677.

Future in Our Hands. 'Join the Movement for a Green Future'. https://www.framtiden.no/en.

Gardiner, Stephen M. 'A Call for a Global Constitutional Convention Focused on Future Generations'. *Ethics & International Affairs* 28, no. 3 (2014): 299–315. https://doi.org/10.1017/S0892679414000379.

Gilabert, Pablo. 'Justice and Feasibility: A Dynamic Approach'. In *Political Utopias: Contemporary Debates*, edited by Kevin Vallier and Michael Weber. Oxford University Press, 2017. https://doi.org/10.1093/acprof:oso/9780190280598.003.0006.

Godfrey-Smith, Peter. *Other Minds: The Octopus, the Sea, and the Deep Origins of Consciousness*. Farrar, Straus and Giroux, 2016.

Godfrey-Smith, Peter. *Living on Earth: Life, Consciousness and the Making of the Natural World*. William Collins, 2025.

Goldsworthy, Lynda. 'Consensus Decision-Making in CCAMLR: Achilles' Heel or Fundamental to Its Success?' *International Environmental Agreements: Politics, Law and Economics* 22, no. 3 (2022): 411–37. https://doi.org/10.1007/s10784-021-09561-4.

Goodin, Robert E. 'Enfranchising the Earth, and Its Alternatives'. *Political Studies* 44, no. 5 (1996): 835–49. https://doi.org/10.1111/j.1467-9248.1996.tb00337.x.

Goodin, Robert E. 'Enfranchising All Affected Interests, and Its Alternatives'. *Philosophy & Public Affairs* 35, no. 1 (2007): 40–68. http://onlinelibrary.wiley.com/doi/10.1111/j.1088-4963.2007.00098.x/full.

Goodin, Robert E., and Gustaf Arrhenius. 'Enfranchising All Subjected: A Reconstruction and Problematization'. *Politics, Philosophy & Economics* 23, no. 2 (2024): 125–53. https://doi.org/10.1177/1470594X241232023.

Gould, Carol C. 'Structuring Global Democracy: Political Communities, Universal Human Rights, and Transnational Representation'. *Metaphilosophy* 40, no. 1 (2009): 24–41. https://doi.org/https://doi.org/10.1111/j.1467-9973.2009.01568.x.

Governing Board of Te Uru Taumatua. 'Governance of Te Urewera'. 2025. https://www.ngaituhoe.iwi.nz/te-urewera-governance.

Graham, Alastair G. C., Anna Wåhlin, Kelly A. Hogan, et al. 'Rapid Retreat of Thwaites Glacier in the Pre-Satellite Era.' *Nature Geoscience* 15, no. 9 (2022): 706–13. https://doi.org/10.1038/s41561-022-01019-9.

Gritsenko, Daria. 'Energy Development in the Arctic: Resource Colonialism Revisited'. In *Handbook of the International Political Economy of Energy and Natural Resources*, edited by Andreas Goldthau, Michael F. Keating, and Caroline Kuzemko. Edward Elgar Publishing, 2018.

Grotius, Hugo. *The Free Sea, With William Welwood's Critique and Grotius's Reply*. Translated by Richard Hakluyt, edited by David Armitage. Liberty Fund, 2004. First published in 1609.

Guerrero, Alexander A. 'Against Elections: The Lottocratic Alternative'. *Philosophy and Public Affairs* 42, no. 2 (2014): 135–78. https://doi.org/10.1111/papa.12029.

Gutmann, Andreas. 'Pachamama as a Legal Person? Rights of Nature and Indigenous Thought in Ecuador'. In *Rights of Nature: A Re-Examination*, edited by Daniel P. Corrigan and Markku Oksanen. Routledge, 2021.

Hayes, J. Gordon. *Antarctica: A Treatise on the Southern Continent.* Richards Press, 1928.

Head, Ivan L. 'Canadian Claims to Territorial Sovereignty in the Arctic Regions'. *McGill Law Journal* 9, no. 3 (1963): 200–26. https://heinonline.org/HOL/P?h=hein.journals/mcgil9&i=206.

Held, David. *Cosmopolitanism: Ideals and Realities.* Polity Press, 2010.

Hemmings, Alan D. 'Considerable Values in Antarctica'. *The Polar Journal* 2, no. 1 (2012): 139–56. http://www.tandfonline.com/doi/abs/10.1080/2154896X.2012.679565.

Hemmings, Alan D. 'Security Beyond Claims'. In *Antarctic Security in the Twenty-First Century: Legal and Policy Perspectives*, edited by Alan D. Hemmings, Donald R. Rothwell, and Karen N. Scott. Routledge, 2012.

Hemmings, Alan D. 'Antarctic Politics in a Transforming Global Geopolitics'. In *Handbook on the Politics of Antarctica*, edited by Klaus Dodds, Alan D. Hemmings, and Peder Roberts. Edward Elgar Publishing, 2017.

Hemmings, Alan D. 'Liability Postponed: The Failure to Bring Annex VI of the Madrid Protocol into Force'. *The Polar Journal* 8, no. 2 (2018): 315–32. https://doi.org/10.1080/2154896X.2018.1541550.

Hemmings, Alan D. 'The Philosophy of Law in the Antarctic'. In *Philosophies of Polar Law*, edited by Dawid Bunikowski and Alan D. Hemmings. Routledge, 2020.

Hemmings, Alan D. 'Realising Rights of Nature Across and Beneath the Southern Ocean'. *The Polar Journal* 14, no. 2 (2024): 409–26. https://doi.org/10.1080/2154896X.2024.2415826.

Hemmings, Alan D., Sanjay Chaturvedi, Elizabeth Leane, Daniela Liggett, and Juan Francisco Salazar. 'Nationalism in Today's Antarctic'. *The Yearbook of Polar Law* VII (2015): 531–55. https://doi.org/10.1163/2211-6427_020.

Hemmings, Alan D., Klaus Dodds, and Peder Roberts. 'Introduction: The Politics of Antarctica'. In *Handbook on the Politics of Antarctica*, edited by Alan D. Hemmings, Klaus Dodds, and Peder Roberts. Edward Elgar Publishing, 2017.

Horta, Oscar. 'Debunking the Idyllic View of Natural Processes: Population Dynamics and Suffering in the Wild'. *Télos* 17, no. 1 (2010): 73–88.

Horta, Oscar. 'Zoopolis, Interventions and the State of Nature'. *Law, Ethics and Philosophy* no. 1 (2013): 113–25. https://raco.cat/index.php/LEAP/article/view/294784.

Howkins, Adrian. 'Appropriating Space: Antarctic Imperialism and the Mentality of Settler Colonialism'. In *Making Settler Colonial Space*, edited by Tracey Banivanua Mar and Penelope Edmonds. Palgrave Macmillan, 2010.

Howkins, Adrian. *Frozen Empires: An Environmental History of the Antarctic Peninsula.* Oxford University Press, 2016.

Howkins, Adrian. *The Polar Regions: An Environmental History.* Polity, 2016.

Hume, David. *A Treatise of Human Nature.* Vol. 2. Everyman's Library, 1974.

Imperial War Museums. 'Why Whales Were Vital in the First World War'. 2025. https://www.iwm.org.uk/history/why-whales-were-vital-in-the-first-world-war.

Innis, Harold A. *The Fur Trade in Canada: An Introduction to Canadian Economic History.* Revised edition. University of Toronto Press, 1970.

International Association of Antarctica Tour Operators. 'IAATO Overview of Antarctic Tourism: the 2022–23 Season, and Preliminary Estimates for 2023–24'. IP 56, ATCM 17, XLV Antarctic Treaty Consultative Meeting, 2023. https://iaato.org/system/files?file=2025-01/ATCM45_ip056_e_IAATO-Overview-of-Antarctic-Tourism-The-2022-23-Season-and-Preliminary-Estimates-for-2023-24.pdf.

International Court of Justice. *Antarctica Cases (United Kingdom v. Argentina; United Kingdom v. Chile).* 1956. https://icj-cij.org/case/26.

International Whaling Commission. 'International Convention for the Regulation of Whaling'. 2 December 1946. https://archive.iwc.int/pages/view.php?ref=3607&k=.

Iorns Magallanes, Catherine J. 'Nature as an Ancestor: Two Examples of Legal Personality for Nature in New Zealand'. *VertigO—La revue électronique en sciences de l'environnement*, hors-série 22 (2015). https://doi.org/10.4000/vertigo.16199.

Isaac, Manuel Gustavo, Steffen Koch, and Ryan Nefdt. 'Conceptual Engineering: A Road Map to Practice'. *Philosophy Compass* 17, no. 10 (2022): 1–15. https://doi.org/https://doi.org/10.1111/phc3.12879.

Jacquet, Jennifer, Eli Blood-Patterson, Cassandra Brooks, and David Ainley. '"Rational Use" in Antarctic Waters'. *Marine Policy* 63 (2016): 28–34. https://doi.org/10.1016/j.marpol.2015.09.031.

Jamieson, Dale. *Ethics and the Environment: An Introduction*. Cambridge University Press, 2008.

Jenouvrier, Stéphanie, Marika Holland, Julienne Stroeve, et al. 'Projected Continent-Wide Declines of the Emperor Penguin Under Climate Change'. *Nature Climate Change* 4, no. 8 (2014): 715–18. https://doi.org/10.1038/nclimate2280.

Jones, Ryan T. *Red Leviathan: The Secret History of Soviet Whaling*. University of Chicago Press, 2022.

Joughin, Ian, Benjamin E. Smith, and Brooke Medley. 'Marine Ice Sheet Collapse Potentially Under Way for the Thwaites Glacier Basin, West Antarctica'. *Science* 344, no. 6185 (2014): 735–38. https://doi.org/10.1126/science.1249055.

Joyner, Christopher C. *Governing the Frozen Commons: The Antarctic Regime and Environmental Protection*. University of South Carolina Press, 1998.

Justus, James. *The Philosophy of Ecology: An Introduction*. Cambridge University Press, 2021.

Kallis, Giorgos, Vasilis Kostakis, Steffen Lange, Barbara Muraca, Susan Paulson, and Matthias Schmelzer. 'Research On Degrowth'. *Annual Review of Environment and Resources* 43, no. 1 (2018): 291–316. https://doi.org/10.1146/annurev-environ-102017-025941.

Keefer, Bowie, Michael Wolovick, and John C. Moore. 'Feasibility of Ice Sheet Conservation Using Seabed Anchored Curtains'. *PNAS Nexus* 2, no. 3 (2023): pgad053. https://doi.org/10.1093/pnasnexus/pgad053.

Khandozhko, Roman. 'Antarctic Minerals for the Soviet Bloc? Imagining the South Pole Frontier of Extractive Socialism'. In *Colonialism and Antarctica: Attitudes, Logics, and Practices*, edited by Peder Roberts and Alejandra Mancilla. Manchester University Press, 2024. https://doi.org/10.7765/9781526170644.00008.

Kohn, Margaret, and Kavita Reddy. 'Colonialism'. In *The Stanford Encyclopedia of Philosophy*, summer 2024 edition, edited by Edward Zalta and Uri Nodelman. https://plato.stanford.edu/archives/sum2024/entries/colonialism/.

Kolers, Avery. *Land, Conflict and Justice*. Cambridge University Press, 2009.

Koskenniemi, Martti. *The Gentle Civilizer of Nations: The Rise and Fall of International Law 1870–1960*. Cambridge University Press, 2001.

Kotzé, Louis J., and Paola Villavicencio Calzadilla. 'Somewhere Between Rhetoric and Reality: Environmental Constitutionalism and the Rights of Nature in Ecuador'. *Transnational Environmental Law* 6, no. 3 (2017): 401–33. https://heinonline.org/HOL/P?h=hein.journals/tevl6&i=421.

Kumar, Krishan. 'Colony and Empire, Colonialism and Imperialism: A Meaningful Distinction?' *Comparative Studies in Society and History* 63, no. 2 (2021): 280–309. https://doi.org/DOI: 10.1017/S0010417521000050.

Kuyper, Jonathan. 'Global Governance and the Democratic Deficit'. *Alternative Law Journal* 35, no. 3 (2010): 169–70.

Lakhtine, W. 'Rights to the Arctic'. *American Journal of International Law* 24, no. 4 (1930): 703–17.

Leopold, Aldo. *A Sand County Almanac, and Sketches Here and There*. Special commemorative edition. Oxford University Press, 1989.

Lewis, R. S. *A Continent for Science: The Antarctic Adventure*. Secker & Warburg, 1965.

List, Christian, and Laura Valentini. 'The Methodology of Political Theory'. In *Oxford Handbook of Philosophical Methodology*, edited by Herman Cappelen, Tamar Szabó Gendler, and John Hawthorne. Oxford University Press, 2016. https://doi.org/10.1093/oxfordhb/9780199668779.013.10.

Liu, Sizhu. 'Ensuring Compliance with Antarctic Environmental Protection Through Inspections: Problems and Prospects'. *Marine Policy* 171 (2025): 106433. https://doi.org/10.1016/j.marpol.2024.106433.

Locke, John. 'Second Treatise of Government'. In *Two Treatises of Government*, edited by Peter Laslett. Cambridge University Press, 1988. First published anonymously in 1689.

Lucci, Juan José, María Alegre, and Leandro Vigna. 'Renewables in Antarctica: An Assessment of Progress to Decarbonize the Energy Matrix of Research Facilities'. *Antarctic Science* 34, no. 5 (2022): 374–88. https://doi.org/DOI: 10.1017/S095410202200030X.

Machowski, Jacek. *The Status of Antarctica in the Light of International Law.* National Center for Scientific, Technical and Economic Information, 1977.

Maddison, Ben. *Class and Colonialism in Antarctic Exploration, 1750–1920.* Pickering & Chatto, 2014.

Magaña, Pablo. 'The Political Representation of Nonhuman Animals'. *Social Theory and Practice* 48, no. 4 (2022): 665–90. https://www.jstor.org/stable/48747320.

Maksimovich, Maya. 'Ukraine Is Going to Compete for Its Leadership in Antarctica'. *News of Ukraine*, 19 January 2019. https://replyua.net/129430-ukraina-sobiraetsya-sopernichat-za-pervenstvo-v-antarktide.html.

Malm, Andreas. *The Progress of This Storm: Nature and Society in a Warming World.* Verso Books, 2018.

Mancilla, Alejandra. 'Rethinking Land and Natural Resources, and Rights over Them'. *Philosophy and Public Issues* 6, no. 2 (2016): 125–41. https://images.fqpjournal.com/f/Archivio_rivista/FQ/FQP_2016_2.

Mancilla, Alejandra. 'Decolonising Antarctica'. In *Philosophies of Polar Law*, edited by Alan Hemmings and Dawid Bunikowski. Routledge, 2021.

Mancilla, Alejandra. 'South American Claims in Antarctica: Colonial, Malgré Tout'. *The Polar Journal* 12, no. 1 (2022): 22–41. https://doi.org/10.1080/2154896X.2022.2062558.

Mancilla, Alejandra. 'Effective Deoccupation: Towards Responsible Guardianship of Nature'. *Political Philosophy* 2, no. 2 (2025). https://doi.org/10.16995/pp.23671.

Mancilla, Alejandra, and Peder Roberts. 'The Antarctic Paradox'. Aeon, 27 May 2022. https://aeon.co/essays/how-do-we-solve-the-paradox-of-protection-in-antarctica.

Manno, C., V. L. Peck, I. Corsi, and E. Bergami. 'Under Pressure: Nanoplastics as a Further Stressor for Sub-Antarctic Pteropods Already Tackling Ocean Acidification'. *Marine Pollution Bulletin* 174 (2022): 113176. https://doi.org/10.1016/j.marpolbul.2021.113176.

Martin-Nielsen, Janet. *A Few Acres of Ice: Environment, Sovereignty and Grandeur in the French Antarctic.* Cornell University Press, 2023. https://doi.org/10.7591/cornell/9781501772092.001.0001.

Mawson, Douglas. *The Home of the Blizzard.* Vol. 1. William Heinemann, 1915.

May, John. *The Greenpeace Book of Antarctica: A New View of the Seventh Continent.* Dorling Kindersley, 1988.

McCord, Edward L. *The Value of Species.* Yale University Press, 2012.

McGee, Jeffrey, David Edmiston, and Marcus G. Haward. *The Future of Antarctica: Scenarios from Classical Geopolitics.* Springer, 2022.

McIvor, Ewan, and Rodolfo Andrés Sánchez. 'The Antarctic Committee for Environmental Protection: Past, Present, and Future'. *Polar Record* 43, no. 3 (2007): 239–46. https://doi.org/DOI: 10.1017/S0032247407006547.

McKitterick, T. E. M. 'The Validity of Territorial and Other Claims in Polar Regions'. *Journal of Comparative Legislation and International Law* 21 (1939): 89–97. https://heinonline.org/HOL/P?h=hein.journals/jclilcs21&i=97.

McMahan, Jeff. 'The Moral Problem of Predation'. In *Philosophy Comes to Dinner: Arguments About the Ethics of Eating*, edited by Andrew Chignell, Terence Cuneo, and Matt Halteman. Routledge: 2015.

McMahon, Kelton W., Chantel I. Michelson, Tom Hart, Matthew D. McCarthy, William P. Patterson, and Michael J. Polito. 'Divergent Trophic Responses of Sympatric Penguin Species to Historic Anthropogenic Exploitation and Recent Climate Change'. *Proceedings of the National Academy of Sciences* 116, no. 51 (2019): 25721–27. https://doi.org/10.1073/pnas.1913093116.

Miller, David. 'Territorial Rights: Concept and Justification'. *Political Studies* 60, no. 2 (2012): 252–68. https://doi.org/10.1111/j.1467-9248.2011.00911.x.

Milligan, Tony. 'The Political Turn in Animal Rights'. *Politics and Animals* 1 (2015): 6–15. https://journals.lub.lu.se/pa/article/view/13512.

Monet, Jenni. '"Green Colonialism": Indigenous World Leaders Warn over West's Climate Strategy'. *The Guardian Online*, 23 April 2023. https://www.theguardian.com/world/2023/apr/23/un-indigenous-peoples-forum-climate-strategy-warning.

Moore, John C., Rupert Gladstone, Thomas Zwinger, and Michael Wolovick. 'Geoengineer Polar Glaciers to Slow Sea-Level Rise'. *Nature* 555, no. 7696 (2018): 303–5. https://doi.org/10.1038/d41586-018-03036-4.

Moore, Margaret. *A Political Theory of Territory*. Oxford University Press, 2015.

Moore, Margaret. 'Justice and Colonialism'. *Philosophy Compass* 11, no. 8 (2016): 447–61. https://doi.org/10.1111/phc3.12337.

Moore, Margaret. 'Is Canada Entitled to the Arctic?' *Canadian Journal of Philosophy* 50, no. 1 (2020): 98–113. https://doi.org/DOI:%252010.1017/can.2019.8.

Morizot, Jean Baptiste. *Manières d'être vivant*. Actes Sud, 2020.

Moutinho, Sofia. 'H5N1 Avian Flu Is Spreading Rapidly in Antarctica'. *Science* (United States) 387, no. 6739 (2025): 1130–31. https://doi.org/10.1126/science.adx3891.

Muntean, William. 'Considerations of Transparency in Antarctica'. Keynote Speech at the Seminar 'Strengthening the Transparency of the Antarctic Treaty System', Japan Society for the Promotion of Science, Antarctic Governance Research, 2025–2030. 31 July 2025. https://youtu.be/apYfbhPJAzU.

National Oceanic and Atmospheric Administration (NOAA). 'M/V Bahia Paraiso; Palmer Station, Antarctica'. IncidentNews, 28 January 1989. https://incidentnews.noaa.gov/incident/6658#!.

National Park Service. 'Roosevelt, Muir, and the Grace of Place'. Last updated 26 August 2021. https://www.nps.gov/yose/learn/historyculture/roosevelt-muir-and-the-grace-of-place.htm.

Nature Editorial. 'Reform the Antarctic Treaty'. *Nature* 558, no. 14 (2018): 161. https://doi.org/10.1038/d41586-018-05368-7.

New Zealand History. 'First Landing on Antarctica'. Updated 21 September 2021. https://nzhistory.govt.nz/media/photo/first-landing-antarctica.

New Zealand History. 'Antarctica and New Zealand'. Updated 2 July 2024. https://nzhistory.govt.nz/politics/antarctica-and-nz/people.

New Zealand Legislation. *Ross Dependency Boundaries and Government Order in Council 1923 (SR 974)*. Reprint 17 August 1923. https://www.legislation.govt.nz/regulation/imperial/1923/0974/latest/DLM1195.html.

New Zealand Public Act no. 7. *Te Awa Tupua (Whanganui River Claims Settlement) Act 2017*. Date of assent 20 March 2017. Version of 27 August 2025. https://www.legislation.govt.nz/act/public/2017/0007/latest/whole.html.

New Zealand Public Act no. 51. *Te Urewera Act* 2014. Date of assent 27 July 2014. Version of 5 April 2025. https://www.legislation.govt.nz/act/public/2014/0051/latest/whole.html.

Nine, Cara. *Global Justice and Territory*. Oxford University Press, 2012.

Nine, Cara. *Sharing Territories: Overlapping Self-Determination and Resource Rights*. Oxford University Press, 2022.

Norsk Polarinstitutt. 'The Fauna of Antarctica'. https://npolar.no/en/themes/fauna-antarctica/.

Nozick, Robert. *Anarchy, State and Utopia*. Basic Books, 1974.

Nussbaum, Martha C. *Justice for Animals: Our Collective Responsibility*. Simon and Schuster, 2022.

Ochoa Espejo, Paulina. 'On Borders: Territories, Legitimacy, and the Rights of Place'. Oxford University Press, 2020.

Odenbaugh, Jay. 'Seeing the Forest and the Trees: Realism about Communities and Ecosystems', *Philosophy of Science* 74, no. 5 (2007): 628–41, https://doi.org/10.1086/525609.

O'Neill, John. 'Representing People, Representing Nature, Representing the World'. *Environment and Planning C: Government and Policy* 19, no. 4 (2001): 483–500. https://doi.org/10.1068/c12s.

O'Reilly, Jessica. *The Technocratic Antarctic: An Ethnography of Scientific Expertise and Environmental Governance*. Cornell University Press, 2017.

Orrego Vicuña, Francisco. 'Antarctic Resources Policy: An Introduction'. In *Antarctic Resources Policy: Scientific, Legal and Political Issues*, edited by Francisco Orrego Vicuña. Cambridge University Press, 1983.

Otto, Ilona M., Jonathan F. Donges, Roger Cremades, et al. 'Social Tipping Dynamics for Stabilizing Earth's Climate by 2050'. *Proceedings of the National Academy of Sciences* 117, no. 5 (2020): 2354–65. https://doi.org/10.1073/pnas.1900577117.

Palmer, Clare. *Animal Ethics in Context*. Columbia University Press, 2010.

Pavel, Carmen E. 'Boundaries, Subjection to Laws, and Affected Interests'. In *The Oxford Handbook of Freedom*, edited by David Schmidtz and Carmen E. Pavel. Oxford University Press, 2018. https://doi.org/10.1093/oxfordhb/9780199989423.013.11.

Pearce, Fred. *The Land Grabbers: The New Fight over Who Owns the Earth*. Beacon Press, 2012.

Phillips, Dom. 'Bolsonaro Declares "The Amazon Is Ours" and Calls Deforestation Data "Lies"'. *The Guardian Online*, 19 July 2019. https://www.theguardian.com/world/2019/jul/19/jair-bolsonaro-brazil-amazon-rainforest-deforestation.

Pinker, Steven. *The Better Angels of Our Nature: The Decline of Violence in History and Its Causes*. Penguin, 2011.

Pinochet de la Barra, Oscar. *Chilean Sovereignty in Antarctica*. Editorial del Pacífico, 1954.

Pinochet de la Barra, Oscar. *Antártica: Sueños del Ayer y del Mañana*. Hachette, 1988.

Pivello, Vânia R., Ima Vieira, Alexander V. Christianini, et al. 'Understanding Brazil's Catastrophic Fires: Causes, Consequences and Policy Needed to Prevent Future Tragedies'. *Perspectives in Ecology and Conservation* 19, no. 3 (2021): 233–55. https://doi.org/https://doi.org/10.1016/j.pecon.2021.06.005.

Plumwood, Val. *Feminism and the Mastery of Nature*. Routledge, 1993.

Plumwood, Val. *Environmental Culture: The Ecological Crisis of Reason*. Routledge, 2002.

Plumwood, Val. 'Meeting the Predator'. In *The Eye of the Crocodile*, edited by Lorraine Shannon. ANU E Press, 2012.

Poe, Edgar Allan. *The Narrative of Arthur Gordon Pym, of Nantucket Comprising the Details of a Mutiny*. Harper & Brothers, 1970.

Powell, Robert B., Stephen R. Kellert, and Sam H. Ham. 'Antarctic Tourists: Ambassadors or Consumers?' *Polar Record* 44, no. 3 (2008): 233–41. https://doi.org/10.1017/S0032247408007456.

Poynter, Charles Wittit. *The Discovery of the South Shetland Islands: The Voyages of the Brig Williams 1819–1820 as Recorded in Contemporary Documents and the Journal of Midshipman C. W. Poynter*, edited by Richard John Campbell. Hakluyt Society, 2000.

Pratt, Angela. 'Treaties vs. Terra Nullius: Reconciliation, Treaty-Making and Indigenous Sovereignty in Australia and Canada'. *Indigenous Law Journal* 3 (2004): 43–60. http://hdl.handle.net/1807/17116.

Prescott, Victor, and Gillian Triggs. *International Frontiers and Borders*. Martinus Nijhoff Publishers, 2008.

Press, Anthony J., Indi Hodgson-Johnston, and Andrew J. Constable. 'The Principles of the Convention on the Conservation of Antarctic Marine Living Resources: Why Its Commission Is Not a Regional Fisheries Management Organization'. In *Governing Marine Living Resources in the Polar Regions*, edited by Nengye Liu, Cassandra M. Brooks, and Tianbao Qin. Edward Elgar Publishing, 2019.

Pyne, Stephen J. *The Ice: A Journey to Antarctica*. University of Iowa Press, 1986.

Quigg, Philip W. *A Pole Apart: The Emerging Issue of Antarctica*. New Press, 1983.

Radío, Pedro. *Soberanía Argentina en la Antártida*. Universidad de Valladolid, 1948.

Ranganathan, Surabhi. 'Global Commons'. *European Journal of International Law* 27, no. 3 (2016): 693–717. https://doi.org/10.1093/ejil/chw037.

Ranganathan, Surabhi. 'Ocean Floor Grab: International Law and the Making of an Extractive Imaginary'. *European Journal of International Law* 30, no. 2 (2019): 573–600. https://doi.org/10.1093/ejil/chz027.

Rapp, Ole Magnus. 'Norge Utvider Dronning Maud Land helt frem til Sydpolen'. *Aftenposten*, 19 September 2015. http://www.aftenposten.no/norge/Norge-utvider-Dronning-Maud-Land-helt-frem-til-Sydpolen-28019b.html.

Rawls, John. *The Law of Peoples*. Harvard University Press, 1999.

Readfearn, Graham. '"Everyone Should Be Concerned": Antarctic Sea Ice Reaches Lowest Levels Ever Recorded'. *The Guardian*, 4 March 2023. https://www.theguardian.com/world/2023/mar/04/everyone-should-be-concerned-antarctic-sea-ice-reaches-lowest-levels-ever-recorded.

Regan, Tom. *The Case for Animal Rights*. University of California Press, 2004.

Remaud, Olivier. *Thinking Like an Iceberg*. Polity Press, 2022.

Republic of Chile. 'Interim Report of the Eleventh Antarctic Treaty Special Consultative Meeting'. Viña del Mar, Chile, 19 November–6 December 1990. https://documents.ats.aq/SATCM11_1/fr/SATCM11_1_fr001_e.pdf.

Richardson, Katherine, Will Steffen, Wolfgang Lucht, et al. 'Earth Beyond Six of Nine Planetary Boundaries'. *Science Advances* 9, no. 37 (2023): eadh2458. https://doi.org/10.1126/sciadv.adh2458.

Ritchie, Hannah, Pablo Rosado, and Max Roser. 'Greenhouse Gas Emissions'. Our World in Data. Last updated January 2024. https://ourworldindata.org/greenhouse-gas-emissions.

Roberts, James. 'Flexibility, Ambiguity and the Construction of the Antarctic Treaty System'. Postgraduate Certificate in Antarctic Studies, University of Canterbury, 2011. https://ir.canterbury.ac.nz/server/api/core/bitstreams/caec9094-4f44-48a3-acdc-545bc3b1b390/content.

Roberts, Leslie Carol. 'The Gigaton Ice Theatre: Performing Ecoactivism in Antarctica'. In *Performing Ice, Performing Landscapes*, edited by Carolyn Philpott, Elizabeth Leane, and Matt Delbridge. Springer International Publishing, 2020. https://doi.org/10.1007/978-3-030-47388-4_9.

Roberts, Peder. *The European Antarctic: Science and Strategy in Scandinavia and the British Empire*. Palgrave Macmillan, 2011.

Roberts, Peder. 'Does the Science Criterion Rest on Thin Ice?' *The Geographical Journal* 189 (2023): 18–24. https://doi.org/10.1111/geoj.12367.

Roberts, Peder, and Alejandra Mancilla, eds. *Colonialism and Antarctica: Attitudes, Logics, and Practices*. Manchester University Press, 2024. https://doi.org/10.7765/9781526170644.

Robinson, Kim Stanley. *Antarctica*. Harper Collins, 1997.

Rocha, Robert C., Phillip J. Clapham, and Yulia V. Ivashchenko. 'Emptying the Oceans: A Summary of Industrial Whaling Catches in the 20th Century'. *Marine Fisheries Review* 76, no. 4 (2014): 37–48.

Rockström, Johan, Louis Kotzé, Svetlana Milutinović, et al. 'The Planetary Commons: A New Paradigm for Safeguarding Earth-Regulating Systems in the Anthropocene'. *Proceedings of the National Academy of Sciences* 121, no. 5 (2024): e2301531121. https://doi.org/10.1073/pnas.2301531121.

Rognhaug, Magnus Hovind. *Norway in the Antarctic*. Norsk Polarinstitutt, 2014.

Rolston III, Holmes. 'Environmental Ethics on Antarctic Ice'. *Polar Record* 36, no. 199 (2000): 289–90. https://doi.org/10.1017/S0032247400016764.

Rolston III, Holmes. 'Environmental Ethics in Antarctica'. *Environmental Ethics* 24, no. 2 (2002): 115–34.

Rolston III, Holmes. *A New Environmental Ethics: The Next Millennium for Life on Earth*. Second edition. Routledge, 2012.

Routley, Richard. 'Is There a Need for a New, an Environmental Ethic?' In *Reflecting on Nature: Readings in Environmental Ethics and Philosophy*. Second edition, edited by Lori Gruen, Dale Jamieson, and Christopher Schlottmann. Oxford University Press, 2013.

Sager, Alex. 'Methodological Nationalism, Migration and Political Theory'. *Political Studies* 64, no. 1 (2014): 42–59. https://doi.org/10.1111/1467-9248.12167.

Sager, Alex. 'Political Philosophy Beyond Methodological Nationalism'. *Philosophy Compass* 16, no. 2 (2021): e12726. https://doi.org/10.1111/phc3.12726.

Sahurie, Emilio J. *The International Law of Antarctica*. Martinus Nijhoff Publishers, 1992.

Saini, Pinki, and Pragya Mishra. 'Bioprospecting of Extremophiles for Industrial Enzymes'. In *Bioprospecting of Microbial Resources for Agriculture, Environment and Bio-Chemical Industry*, edited by Pinki Saini and Pragya Mishra. Springer Nature Switzerland, 2024. https://doi.org/10.1007/978-3-031-63844-2_3.

Sandven, Hallvard. 'Feasibility: Superfluous or Moralized'. *Ethics* 136, no. 1 (2025): 88–120. https://doi.org/10.1086/736744.

Saunders, Ben. 'Defining the Demos'. *Politics, Philosophy & Economics* 11, no. 3 (2012): 280–301. https://doi.org/10.1177/1470594X11416782.

Scambos, Ted A., Robin E. Bell, Richard B. Alley, et al. 'How Much, How Fast?: A Science Review and Outlook for Research on the Instability of Antarctica's Thwaites Glacier in the 21st Century'. *Global and Planetary Change* 153 (2017): 16–34.

Schleicher, Judith, Julie G. Zaehringer, Constance Fastré, Bhaskar Vira, Piero Visconti, and Chris Sandbrook. 'Protecting Half of the Planet Could Directly Affect over One Billion People'. *Nature Sustainability* 2, no. 12 (2019): 1094–96.

Schmidt Prado, Hugo. *¡Base O'Higgins Sin Novedad!* Second edition. Editorial La Noria, 1992.

Schrijver, Nico. *Sovereignty over Natural Resources: Balancing Rights and Duties*. Cambridge University Press, 1997.

Schwartz, Avshalom M. 'Political Imagination and Its Limits'. *Synthese* 199, no. 1 (2021): 3325–43. https://doi.org/10.1007/s11229-020-02936-1.

Schwerdtfeger, Peter. 'On Icebergs and Their Uses'. Report to the Australian Academy of Sciences, 1978.

Scientific Committee on Antarctic Research. 'The 1st SCAR Antarctic and Southern Ocean Science Horizon Scan—Final List of Questions'. *Nature* (Supplementary Information) 512 (2014): 2–5. https://www.nature.com/news/polopoly_fs/7.18965.1407332749!/suppinfoFile/512023a_s1.pdf.

Scott, Shirley V. 'Ingenious and Innocuous? Article IV of the Antarctic Treaty as Imperialism'. *The Polar Journal* 1, no. 1 (2011): 51–62.

Scott, Shirley V. 'Imperialism in Antarctica'. In *The Palgrave Encyclopedia of Imperialism and Anti-Imperialism*, edited by Immanuel Nez and Zak Cope. Springer, 2021. https://doi.org/10.1007/978-3-319-91206-6_142-1.

Searle, John. *The Construction of Social Reality*. The Free Press, 1995.

Secretariat of the Antarctic Treaty. 'Final Report of the Thirty-Ninth Antarctic Treaty Consultative Meeting (Santiago de Chile, 23 May–1 June 2016)'. Vols. 1 & 2. Buenos Aires, 2016. https://documents.ats.aq/ATCM39/fr/ATCM39_fr001_e.pdf and https://documents.ats.aq/ATCM39/fr/ATCM39_fr002_e.pdf.

Secretariat of the Antarctic Treaty. 'Guidelines on the Procedure to be Followed with Respect to Consultative Party Status (Decision 2, ATCM XL—CEP XX, Beijing)'. Adopted 1 June 2017. https://www.ats.aq/devAS/Meetings/Measure/653.

Secretariat of the Antarctic Treaty. 'Parties'. 2025. https://www.ats.aq/devAS/Parties?lang=e.

Seed, Patricia. *Ceremonies of Possession in Europe's Conquest of the New World 1492–1640*. Cambridge University Press, 1995.

Sengupta, Somini, Catrin Einhorn, and Manuela Andreoni. 'There's a Global Plan to Conserve Nature: Indigenous People Could Lead the Way'. *The New York Times*, 23 June 2023. https://www.nytimes.com/2021/03/11/climate/nature-conservation-30-percent.html?searchResultPosition=1.

Shapley, Deborah. *The Seventh Continent: Antarctica in a Resource Age*. Resources for the Future, 1985.

Shaw, Justine D., Aleks Terauds, Martin J. Riddle, Hugh P. Possingham, and Steven L. Chown. 'Antarctica's Protected Areas Are Inadequate, Unrepresentative, and at Risk'. *PLOS Biology* 12, no. 6 (2014): 1–5. https://doi.org/10.1371/journal.pbio.1001888.

Shaw, Malcolm N. *International Law*. Eighth edition. Cambridge University Press, 2017.

Sidder, Aaron. 'Antarctica Could Lose Most of Its Penguins to Climate Change'. *National Geographic*, 29 June 2016. https://www.nationalgeographic.com/animals/article/adelie-penguins-antarctica-climate-change-population-decline-refugia.

Simmons, A. John. *The Lockean Theory of Rights*. Princeton University Press, 1992.

Simmons, A. John. 'On the Territorial Rights of States'. *Philosophical Issues* 11 (2001): 300–26. https://doi.org/10.1111/0029-4624.35.s1.12.

Singer, Peter. *Animal Liberation*. Harper Collins, 2002.

Smedal, Gustav. 'Acquisition of Sovereignty over Polar Areas'. Translated by Chr. Meyer. In *Skrifter Om Svalbard og Ishavet no. 36*. I kommisjon hos Jacob Dybwad, 1931.

Smith, Adam. 'The Theory of Moral Sentiments'. In *The Glasgow Edition of the Works and Correspondence of Adam Smith*, vol. 1, edited by A. L. Macfie and D. D. Raphael. Liberty Fund, 1982.

Sobrevila, Claudia. *The Role of Indigenous Peoples in Biodiversity Conservation: The Natural But Often Forgotten Partners*. World Bank, 2008. https://documents1.worldbank.org/curated/en/995271468177530126/pdf/443000WP0BOX321onservation01PUBLIC1.pdf.

Soga, Masashi, and Kevin J. Gaston. 'Shifting Baseline Syndrome: Causes, Consequences, and Implications'. *Frontiers in Ecology and the Environment* 16, no. 4 (2018): 222–30. https://doi.org/10.1002/fee.1794.

Sollie, Finn. 'The Political Experiment in Antarctica'. In *Frozen Future*, edited by Richard S. Lewis and Philip M. Smith. Quadrangle Books, 1973.

Song, Sarah. 'The Boundary Problem in Democratic Theory: Why the Demos Should Be Bounded by the State'. *International Theory* 4, no. 1 (2012): 39–68. https://doi.org/10.1017/S1752971911000248.

Southwood, Nicholas. 'The Feasibility Issue'. *Philosophy Compass* 13, no. 8 (2018): e12509. https://doi.org/10.1111/phc3.12509.

Stanley, Henry Morton. *Through the Dark Continent: Or, the Sources of the Nile, Around the Great Lakes of Equatorial Africa and Down the Livingstone River to the Atlantic Ocean*. Vol. 2. William Clowes and Sons, 1878. https://archive.org/details/throughdarkcont04stangoog.

Stefansson, Vilhjalmur. *The Friendly Arctic*. The Macmillan Company, 1922.

Stilz, Anna. 'Why Do States Have Territorial Rights?' *International Theory* 1, no. 2 (2009): 185–213. https://doi.org/10.1017/S1752971909000104.

Stilz, Anna. 'Decolonization and Self-Determination'. *Social Philosophy and Policy* 32, no. 1 (2015): 1–24. https://doi.org/DOI: 10.1017/S0265052515000059.

Stilz, Anna. *Territorial Sovereignty: A Philosophical Exploration*. Oxford University Press, 2019.

Stokstad, Erik. 'This Lagoon Is Effectively a Person, Says Spanish Law That's Attempting to Save It'. *Science* 378, no. 6615 (2022): 15–16. https://www.science.org/content/article/lagoon-effectively-person-says-spanish-law-s-attempting-save-it.

Stone, Christopher. 'Should Trees Have Standing? Toward Legal Rights for Natural Objects'. *Southern California Law Review* 45, no. 450 (1972): 450–501.

Stone, Christopher. *Should Trees Have Standing? Law, Morality and the Environment*. Third edition. Oxford University Press, 2010.

Tamm, Sune. 'Peace vs. Compliance in Antarctica: Inspections and the Environment'. *The Polar Journal* 8, no. 2 (2018): 333–50. https://doi.org/10.1080/2154896X.2018.1541551.

Tanasescu, Mihnea. 'Rethinking Representation: The Challenge of Non-Humans'. *Australian Journal of Political Science* 49, no. 1 (2014): 40–53. https://doi.org/10.1080/10361146.2013.874404.

Tarroux, Arnaud, Per Fauchald, and Øystein Varpe. 'An Ocean of Possibilities: Antarctic Petrels' Foraging in the Southern Ocean'. The Fram Centre. Published 18 January 2021. https://framsenteret.no/forum/fram-forum-2021/an-ocean-of-possibilities-antarctic-petrels-foraging-in-the-southern-ocean/.

Tin, Tina, Z. L. Fleming, K. A. Hughes, et al. 'Impacts of Local Human Activities on the Antarctic Environment'. *Antarctic Science* 21, no. 1 (2009): 3–33. http://journals.cambridge.org/abstract_S0954102009001722.

Tønnessen, Johan Nicolay, and Arne Odd Johnsen. *The History of Modern Whaling*. C. Hurst & Company and Australian National University Press, 1982.

Trathan, Philip N., Barbara Wienecke, Christophe Barbraud, et al. 'The Emperor Penguin—Vulnerable to Projected Rates of Warming and Sea Ice Loss'. *Biological Conservation* 241 (2020): 108216. https://doi.org/10.1016/j.biocon.2019.108216.

'Treaty Between Norway, the United States of America, Denmark, France, Italy, Japan, the Netherlands, Great Britain and Ireland and the British Overseas Dominions and Sweden Concerning Spitsbergen Signed in Paris 9th February 1920'. Arctic Portal. http://library.arcticportal.org/1909/1/The_Svalbard_Treaty_9ssFy.pdf.

UNESCO. 'Convention for the Protection of the World Cultural and Natural Heritage'. 16 November 1972. https://unesdoc.unesco.org/ark:/48223/pf0000002091.

United Kingdom, Australia, Belgium, et al., 'Antarctica in a Changing Climate'. WP32, ATCM 16, ATCM XLIII—CEP XXIII, 29 April 2021. https://documents.ats.aq/ATCM43/wp/ATCM43_wp032_e.docx.

United Nations. 'United Nations Framework Convention on Climate Change'. 1992. https://unfccc.int/resource/docs/convkp/conveng.pdf.

United Nations. 'Paris Agreement'. 2015. https://unfccc.int/sites/default/files/english_paris_agreement.pdf.

United Nations Environment Programme (Secretariat of the Convention on Biological Diversity). 'Convention on Biological Diversity'. Entered in force 29 December 1993. Published 2011. https://www.cbd.int/doc/legal/cbd-en.pdf.

United Nations General Assembly. 'Permanent Sovereignty over Natural Resources'. General Assembly Resolution 1803 (XVII), 1962. https://docs.un.org/en/a/res/1803(xvii).

United Nations General Assembly. 'Official Records, 29 September 1982'. A/37/PV.10.

United Nations General Assembly. 'Official Records, 28 November 1984'. A/39/PV.50.

U.S. Fish and Wildlife Service. 'U.S. Fish and Wildlife Service Provides Endangered Species Act Protections for Emperor Penguin'. 25 October 2022. https://www.fws.gov/press-release/2022-10/emperor-penguin-gets-endangered-species-act-protections.

Valentini, Laura. 'Ideal vs. Non-Ideal Theory: A Conceptual Map'. *Philosophy Compass* 7, no. 9 (2012): 654–64. https://doi.org/10.1111/j.1747-9991.2012.00500.x.

Valentini, Laura. 'On the Distinctive Procedural Wrong of Colonialism'. *Philosophy & Public Affairs* 43, no. 4 (2015): 312–31. https://doi.org/10.1111/papa.12057.

Vatn, Arild, and Paul Vedeld. 'Fit, Interplay, and Scale: A Diagnosis'. *Ecology and Society* 17, no. 4 (2012). http://dx.doi.org/10.5751/ES-05022-170412.

Vecchiato, Marco, Elena Argiriadis, Stefano Zambon, et al. 'Persistent Organic Pollutants (POPs) in Antarctica: Occurrence in Continental and Coastal Surface Snow'. *Microchemical Journal* 119 (2015): 75–82. https://doi.org/https://doi.org/10.1016/j.microc.2014.10.010.

Villavicencio Calzadilla, Paola, and Louis J. Kotzé. 'Living in Harmony with Nature? A Critical Appraisal of the Rights of Mother Earth in Bolivia'. *Transnational Environmental Law* 7, no. 3 (2018): 397–424. https://doi.org/10.1017/S2047102518000201.

Wadiwel, Dinesh Joseph. 'Do Fish Resist?' *Cultural Studies Review* 22, no. 1 (2016): 196–242. https://doi.org/10.5130/csr.v22i1.4363.

Waldron, Jeremy. 'Superseding Historic Injustice'. *Ethics* 103, no. 1 (1992): 4–28.

Walters, Tiara. 'Using Cape Town as a Launchpad, Russia Boasts of Supergiant Oil Fields in Antarctic Wilderness'. *Daily Maverick*, 25 October 2021. https://www.dailymaverick.co.za/article/2021-10-25-using-cape-town-as-a-launchpad-russia-boasts-of-supergiant-oil-fields-in-antarctic-wilderness/.

Walters, Tiara. '"Krill, Baby, Krill"—What President-Elect Donald Trump Means for Imperilled Antarctica'. *Daily Maverick*, 25 November 2024. https://www.dailymaverick.co.za/article/2024-11-25-krill-baby-krill-what-president-elect-donald-trump-means-for-imperilled-antarctica/.

Watt-Cloutier, Sheila. *The Right to be Cold: One Woman's Fight to Protect the Arctic and Save the Planet from Climate Change*. University of Minnesota Press, 2018. https://doi.org/10.5749/j.ctt2204r9f.

Weeks, W. F. 'Iceberg Water: An Assessment'. In *Proceedings of the Conference on Use of Icebergs: Scientific and Practical Feasibility, Cambridge, UK, 1–3 April 1980*. International Glaciological Society, 1980.

Wehi, Priscilla M., Nigel J. Scott, Jacinta Beckwith, et al. 'A Short Scan of Māori Journeys to Antarctica'. *Journal of the Royal Society of New Zealand* 52, no. 5 (2022): 587–98. https://doi.org/10.1080/03036758.2021.1917633.

Wehi, Priscilla M., Vincent van Uitregt, Nigel J. Scott, et al. 'Transforming Antarctic Management and Policy with an Indigenous Māori Lens.' *Nature Ecology & Evolution* 5 (2021): 1055–59. https://doi.org/10.1038/s41559-021-01466-4.

Wenar, Leif. *Blood Oil: Tyrants, Violence and the Rules That Run the World*. Oxford University Press, 2016.

Wesche, Philipp. 'Rights of Nature in Practice: A Case Study on the Impacts of the Colombian Atrato River Decision.' *Journal of Environmental Law* 33, no. 3 (2021): 531–55. https://doi.org/10.1093/jel/eqab021.

Whelan, Frederick G. 'Prologue: Democratic Theory and the Boundary Problem.' *Nomos XXV: Liberal Democracy* 25 (1983): 13–47.

Williams, Bernard. *Ethics and the Limits of Philosophy*. Fontana Press/Collins, 1985.

Williams, Bernard. *In the Beginning Was the Deed: Realism and Moralism in Political Argument*. Princeton University Press, 2005.

Wilson, Edward O. *Half-Earth: Our Planet's Fight for Life*. W. W. Norton & Company, 2016.

Wilson, Robert E. 'National Interests and Claims in the Antarctic'. *ARCTIC* 17, no. 1 (1964): 15–31. http://arctic.synergiesprairies.ca/arctic/index.php/arctic/article/view/3481.

Wimmer, Andreas, and Nina Glick Schiller. 'Methodological Nationalism and Beyond: Nation-State Building, Migration and the Social Sciences'. *Global Networks* 2, no. 4 (2002): 301–34.

Winter, Kate. 'Antarctica's First Zero Emission Research Station Shows That Sustainable Living Is Possible Anywhere'. *The Conversation*, 19 November 2019. https://theconversation.com/antarcticas-first-zero-emission-research-station-shows-that-sustainable-living-is-possible-anywhere-113977.

Wolfrum, Rüdiger. 'Liability for Environmental Damage in Antarctica: Supplement to the Rules on State Responsibility or a Lost Opportunity?' In *International Law Between Universalism and Fragmentation*, edited by Isabelle Buffard, James Crawford, Alan Pellet, and Stephan Wittich. Martinus Nijhoff Publishers, 2008. https://doi.org/10.1163/ej.9789004167278.v-1086.264.

Woods, Mark. *Rethinking Wilderness.* Broadview Press, 2017.

Yermakova, Yelena. 'Legitimacy of the Antarctic Treaty System: Is It Time for a Reform?' *The Polar Journal* 11 (2021): 1–18. https://doi.org/10.1080/2154896X.2021.1977048.

Yermakova, Yelena. 'Domination as a Legacy of the Colonial Origins and Structure of the Antarctic Treaty System'. In *Colonialism and Antarctica: Attitudes, Logics, and Practices*, edited by Peder Roberts and Alejandra Mancilla. Manchester University Press, 2024. https://doi.org/10.7765/9781526170644.00018.

Young, Oran R. 'Environmental Governance: The Role of Institutions in Causing and Confronting Environmental Problems'. *International Environmental Agreements* 3 (2003): 377–93.

Young, Oran R. 'Institutions and Environmental Change: The Scientific Legacy of a Decade of IDGEC Research'. In *Institutions and Environmental Change*, edited by Oran R. Young, Leslie A. King, and Heike Schroeder. MIT Press, 2008.

Ypi, Lea. 'What's Wrong with Colonialism'. *Philosophy & Public Affairs* 41, no. 2 (2013): 158–91. https://doi.org/10.1111/papa.12014.

Zielinski, Sarah. 'New Water Bear Species Found in Antarctica'. *ScienceNews*, 13 July 2014. https://www.sciencenews.org/article/new-water-bear-species-found-antarctica.

Index

For the benefit of digital users, indexed terms that span two pages (e.g., 52–53) may, on occasion, appear on only one of those pages.

Figures and tables are indicated by an italic *f* or *t*.